MATH MONTESSORI MATERIALS

Fy Llyfr Gwaith Mathe-mateg Cyntaf

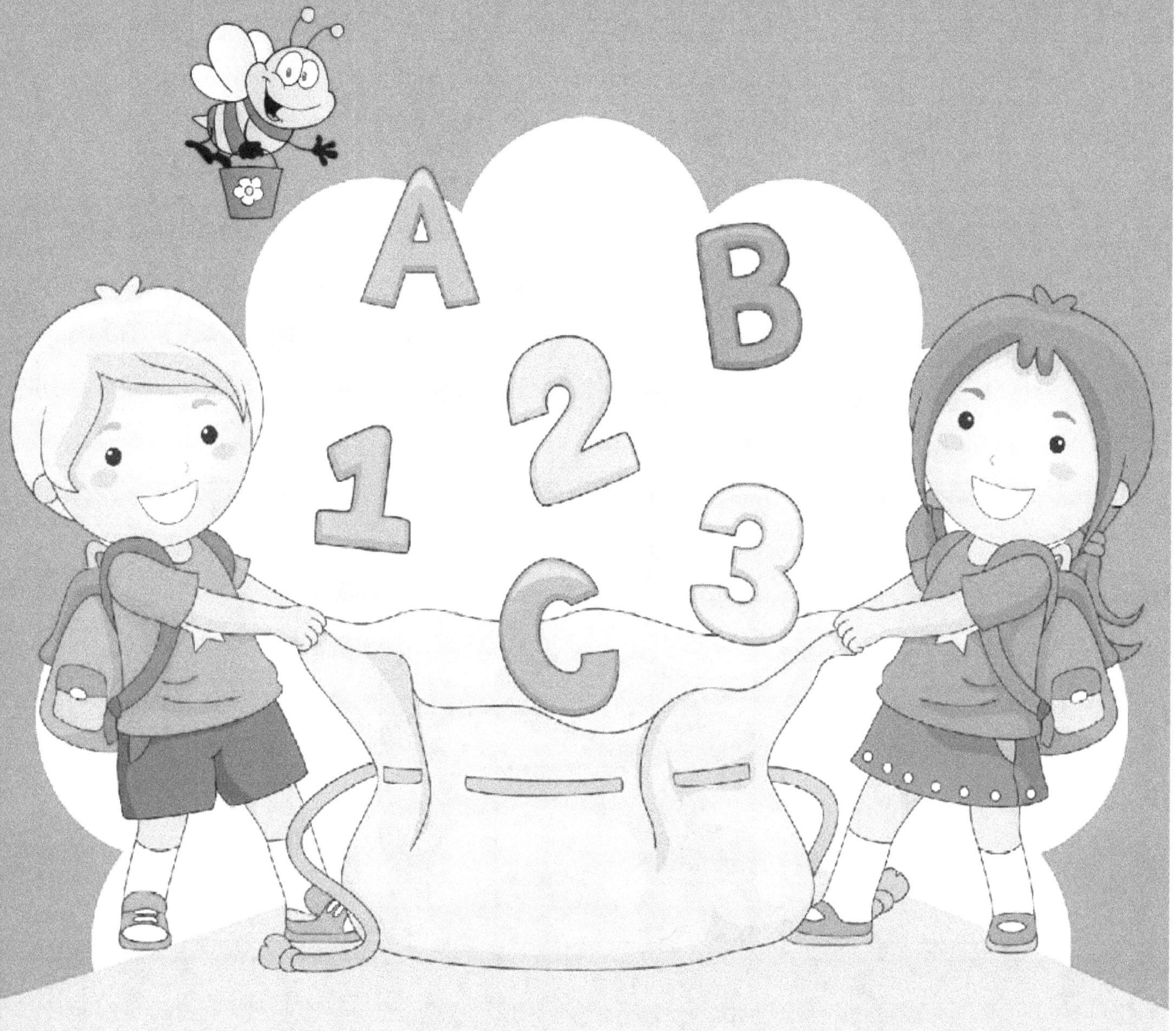

Addition Worksheets

2 + 2	9 + 10	8 + 7
7 + 8	2 + 10	5 + 5
4 + 5	5 + 1	2 + 1

Math Made Easy....

Name : _______________

Direction: Count the images. Write the number of images in the boxes above each image and write the total number in the last box.

+ =

+ =

+ =

+ =

Name : ______________________

Direction: Add the number of images in each box and write the answer in the last box.

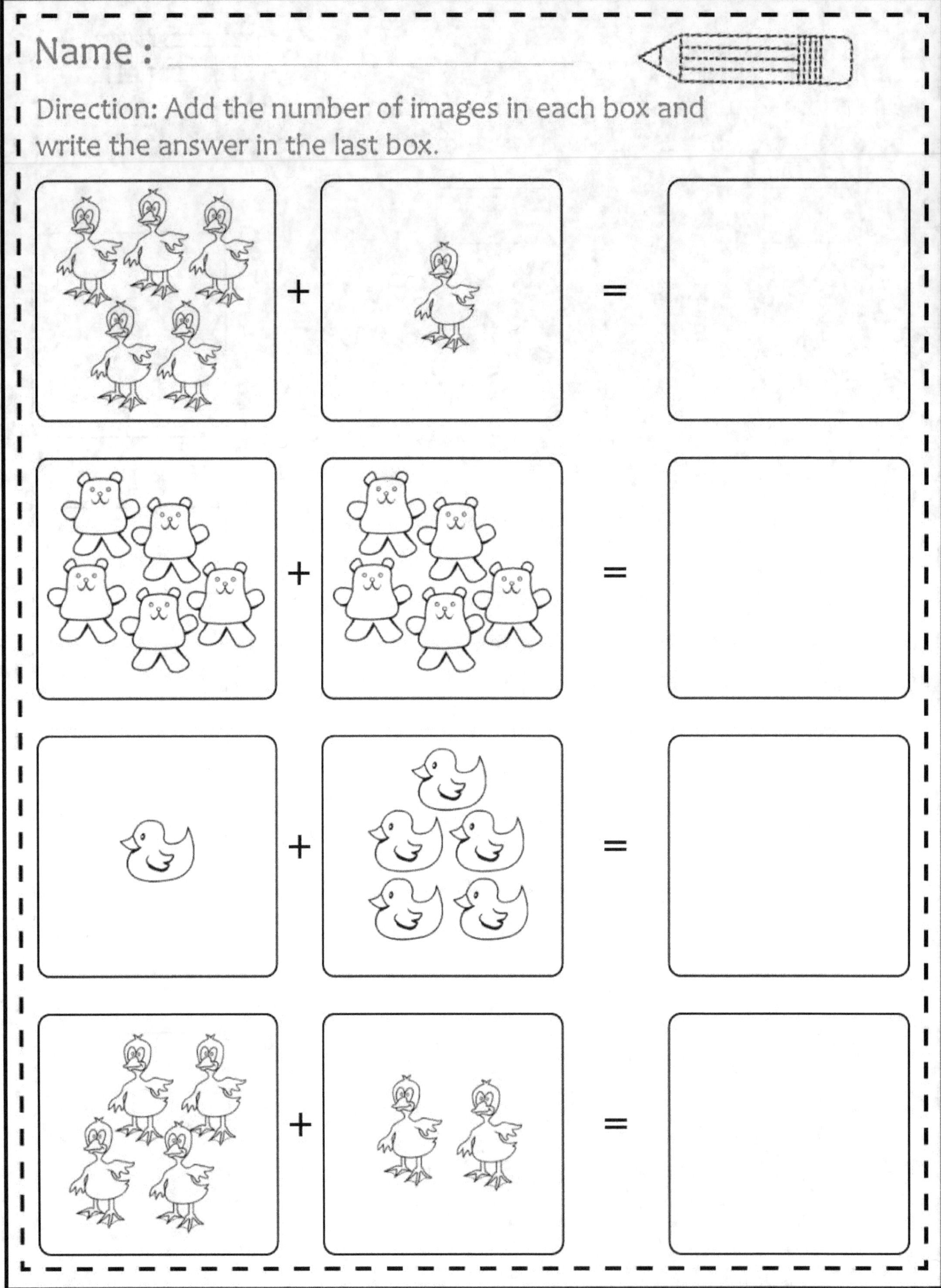

Name : ___________________

Addition Worksheets

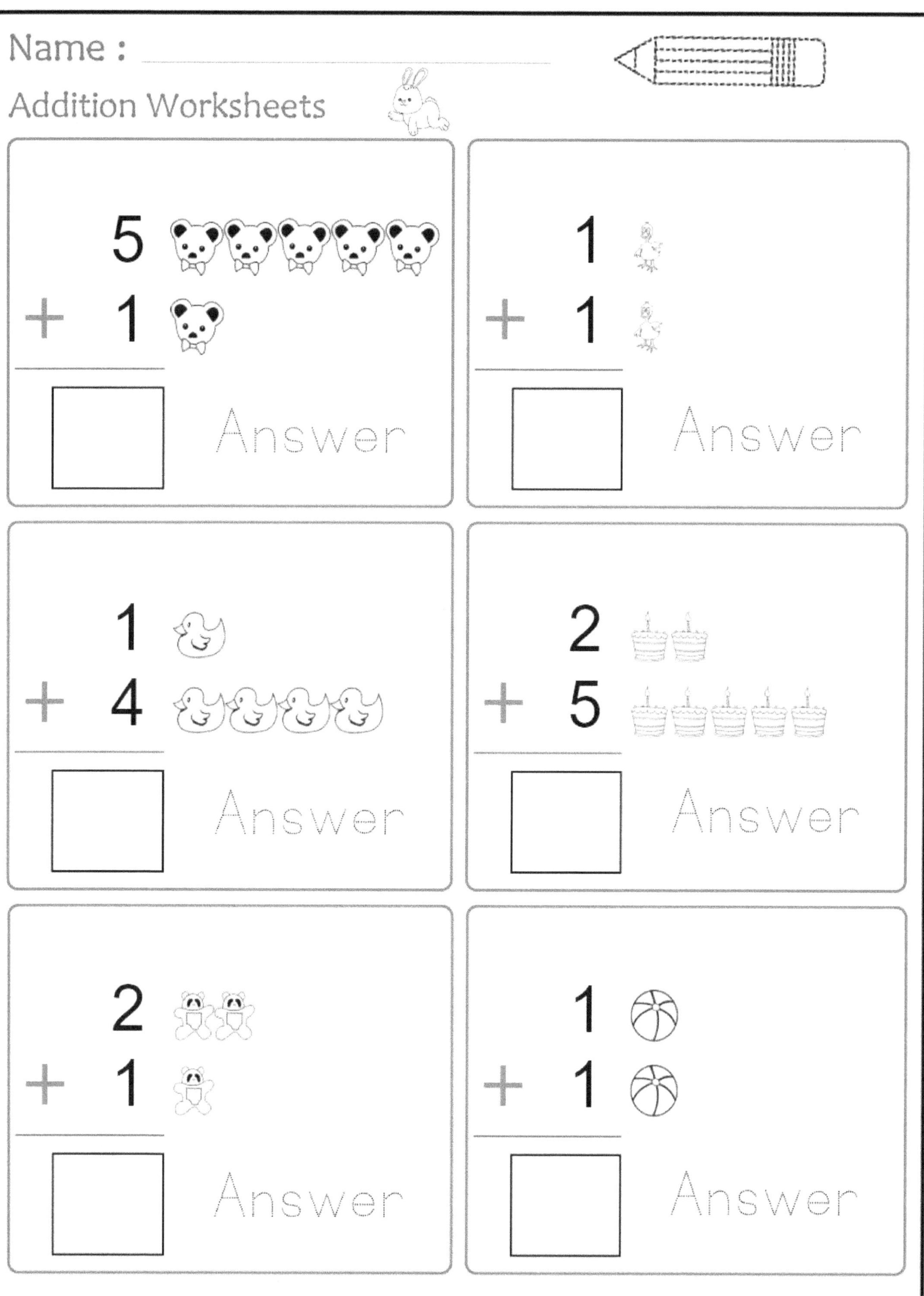

Name : ____________________

Direction: Add the images. Draw a line between the total number of images in each box and the number on the right.

2

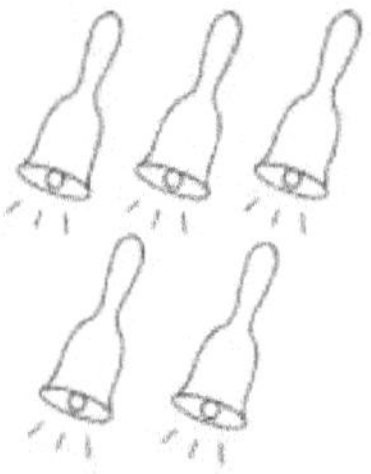

9

9

3

Name : _______________

Addition Worksheets

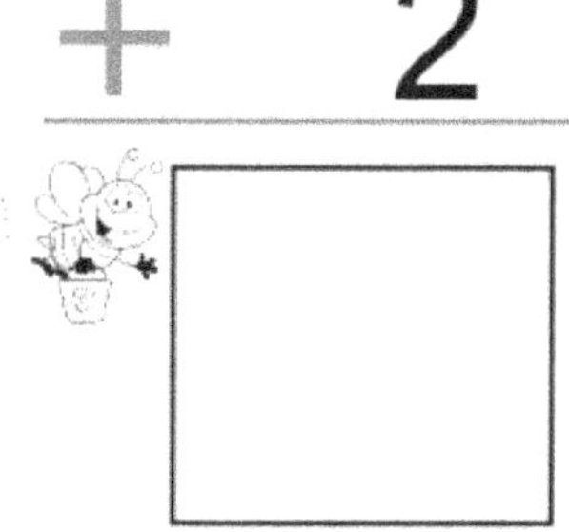

$$4 + 2$$

$$3 + 10$$

$$7 + 7$$

$$7 + 4$$

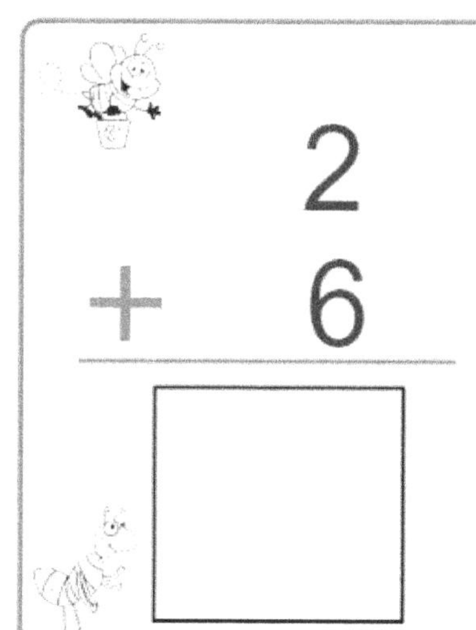

$$2 + 6$$

$$3 + 10$$

$$10 + 9$$

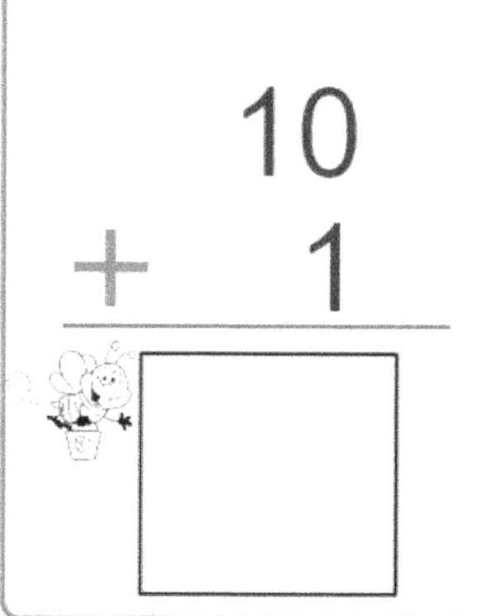

$$10 + 1$$

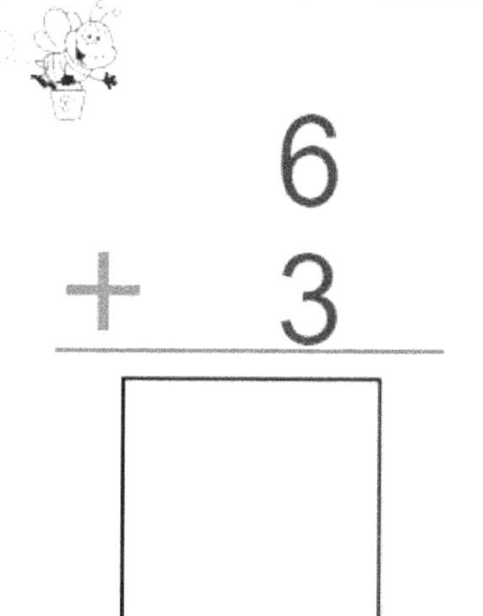

$$6 + 3$$

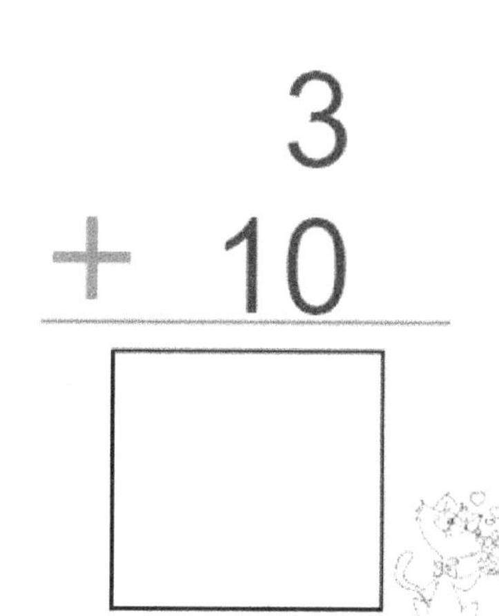

Math Made Easy....

Name : ______________________________

Direction: Count the images. Write the number of images in the
boxes above each image and write the total number in the last box.

Name : __________________________

Direction: Add the number of images in each box and
write the answer in the last box.

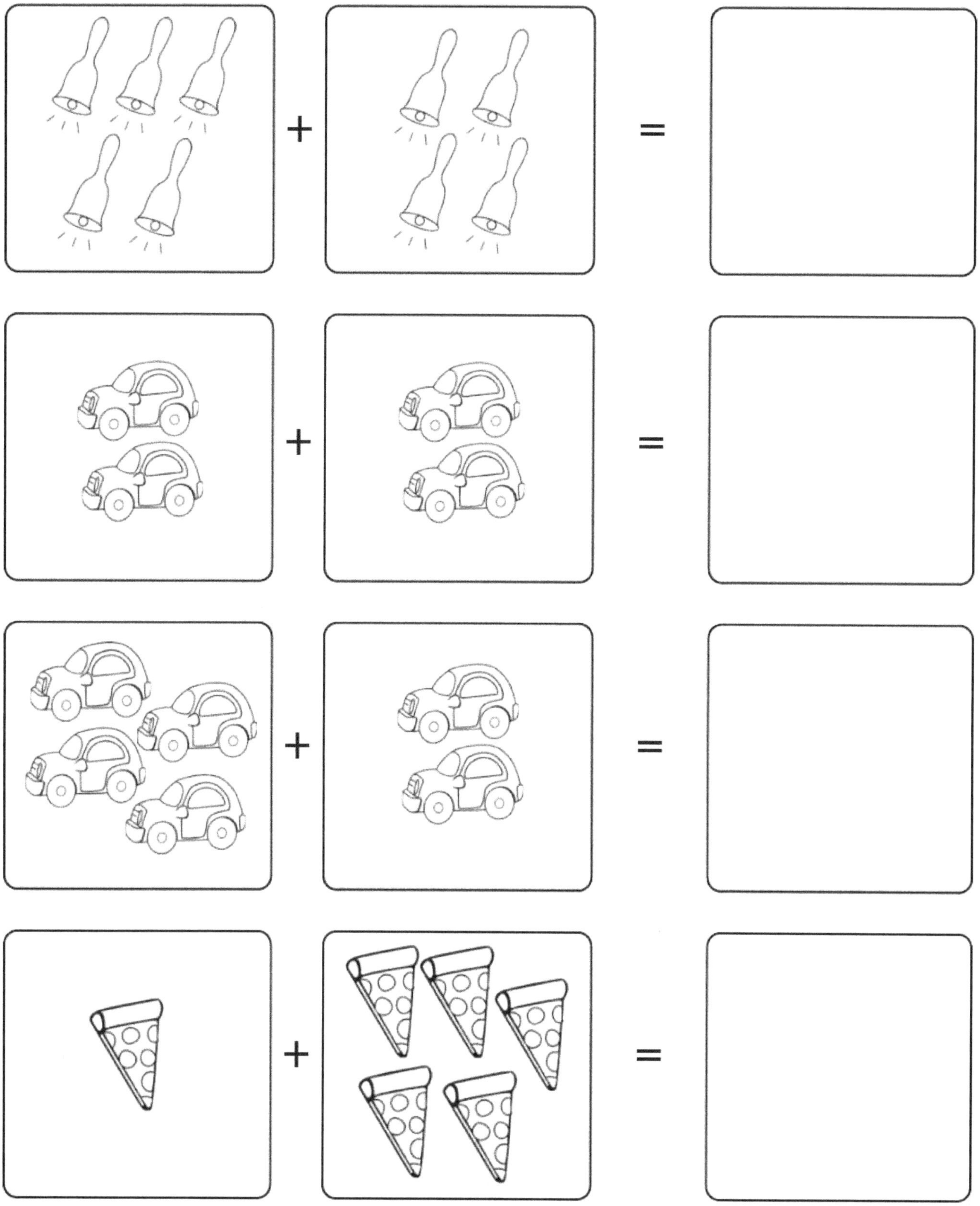

Name : _______________________

Addition Worksheets

1 
+ 5

☐ Answer

1 
+ 3

☐ Answer

4 
+ 5

☐ Answer

4 
+ 1

☐ Answer

2 
+ 1

☐ Answer

4 
+ 3

☐ Answer

Name : _______________________

Direction: Add the images. Draw a line between the total
number of images in each box and the number on the right.

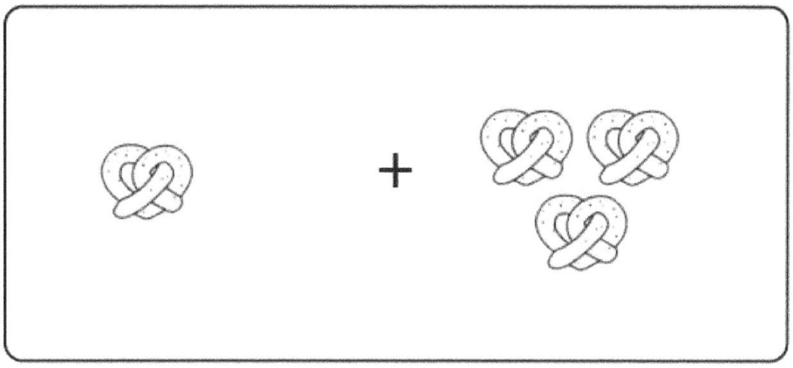

4

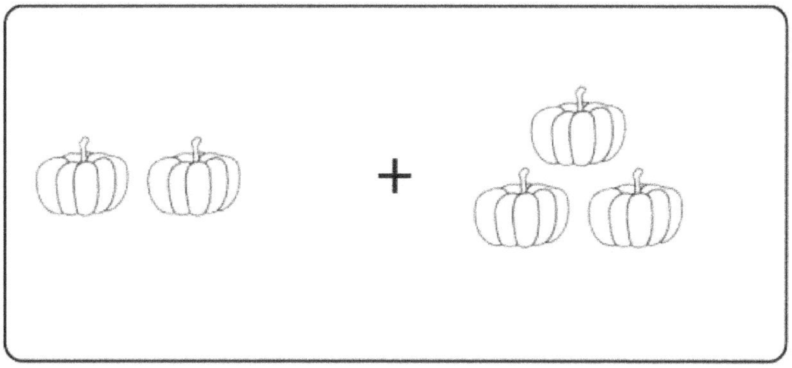

8

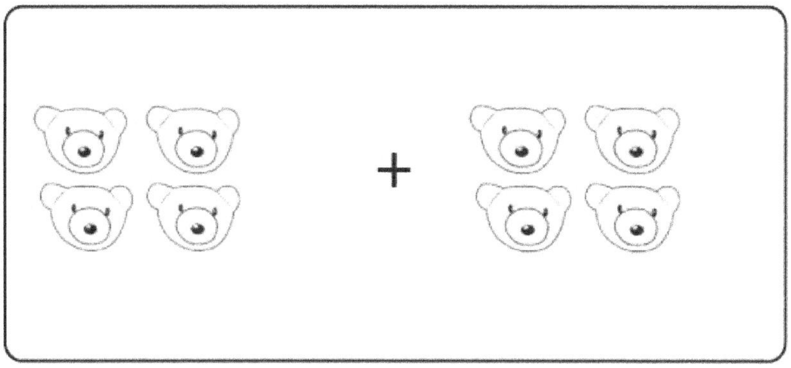

8

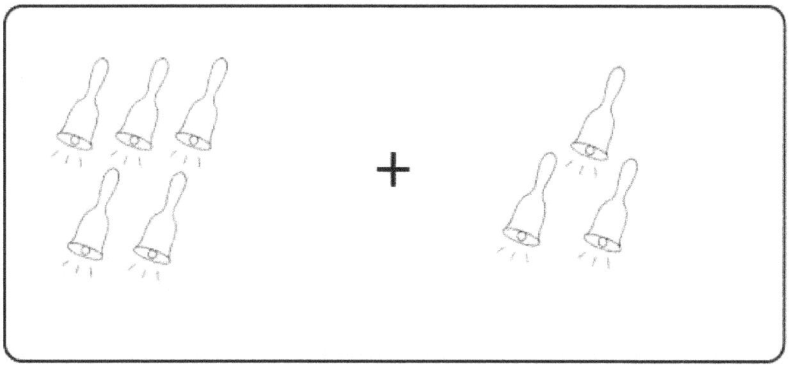

5

Addition Worksheets

```
    6          9          3
  + 1        + 2        + 7
  -----      -----      -----
```
 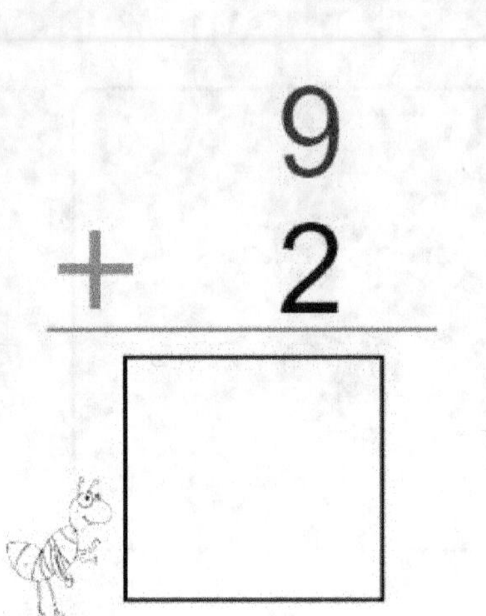 

```
    7          7         10
  + 6        + 5        + 7
  -----      -----      -----
```
 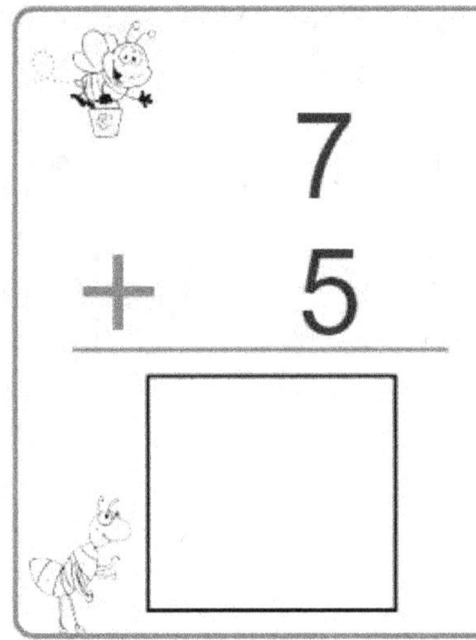 

```
    3          6         10
  + 8        + 5        + 2
  -----      -----      -----
```
  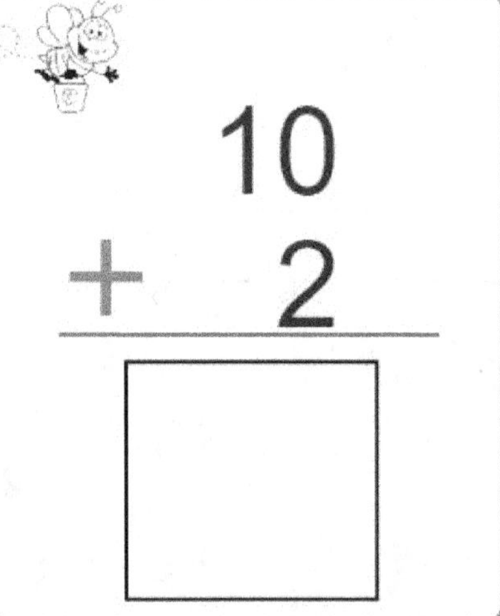

 Math Made Easy.... 

Name : _______________________

Direction: Count the images. Write the number of images in the
boxes above each image and write the total number in the last box.

+ =

+ =

+ =

+ =

Name : _________________________

Direction: Add the number of images in each box and write the answer in the last box.

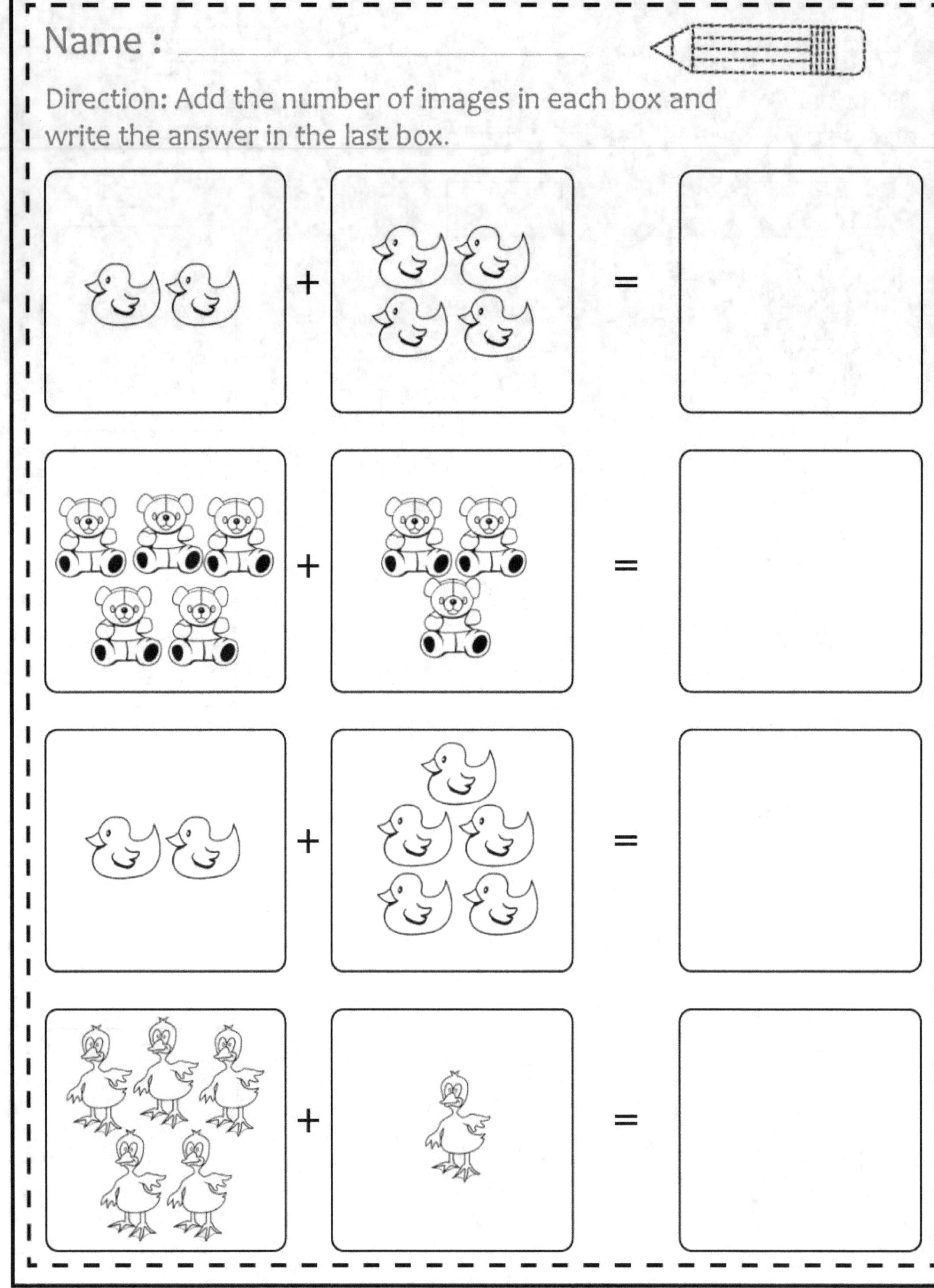

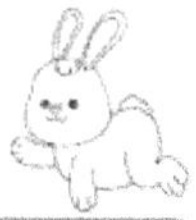

1
+ 5

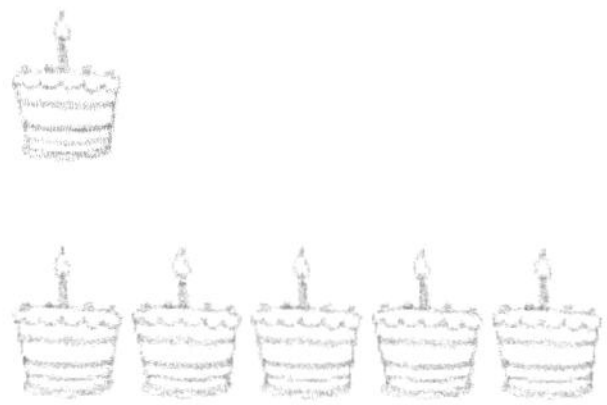

Answer

2
+ 5

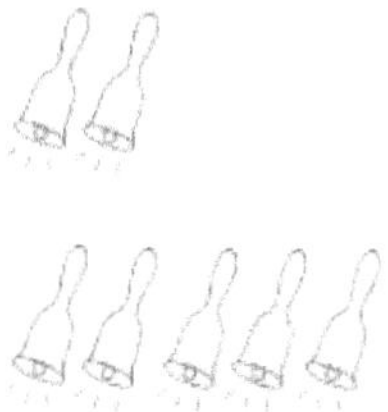

Answer

2
+ 2

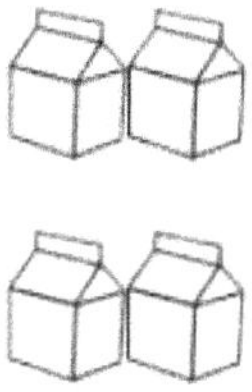

Answer

5
+ 3

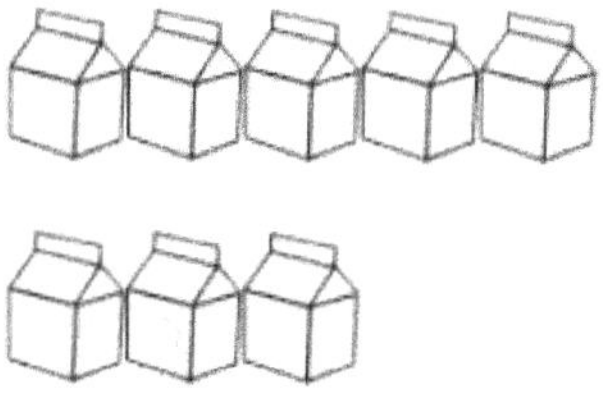

Answer

5
+ 2

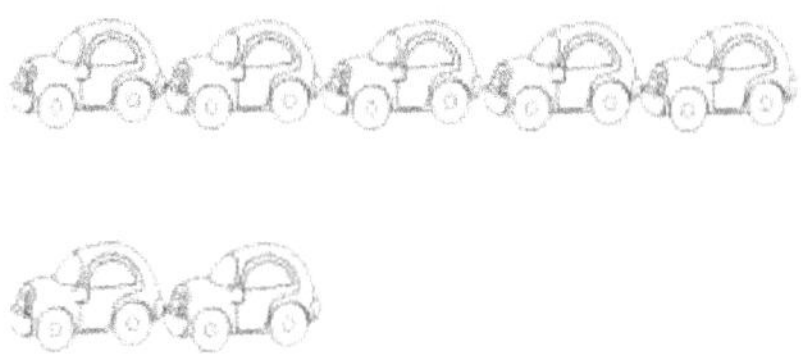

Answer

1
+ 2

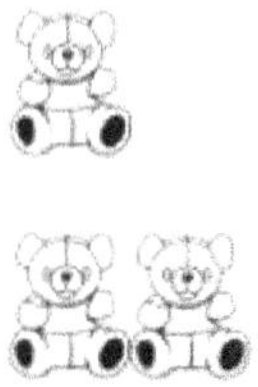

Answer

Direction: Add the images. Draw a line between the total number of images in each box and the number on the right.

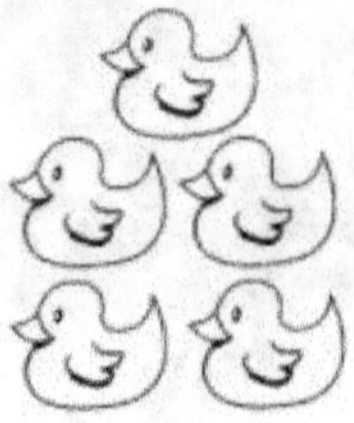

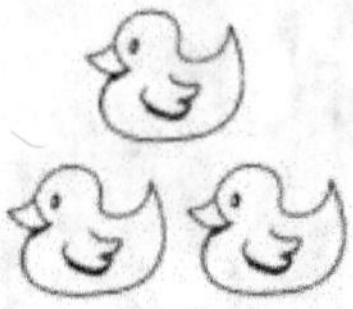

9

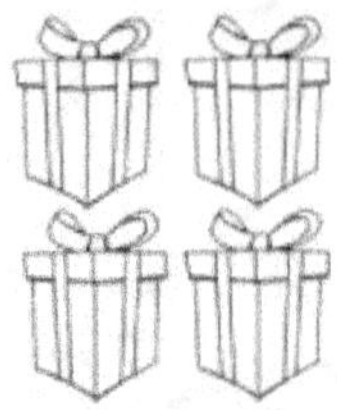

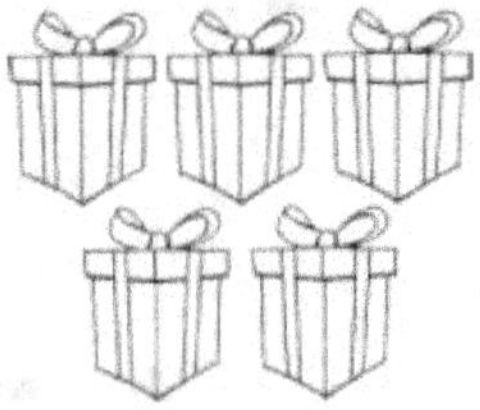

4

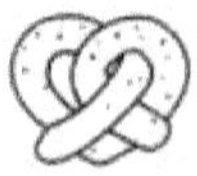

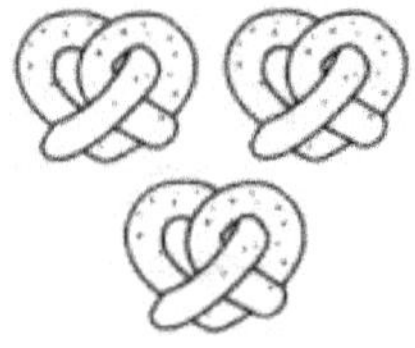

9

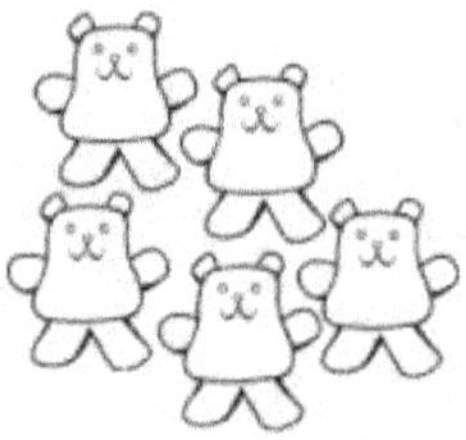

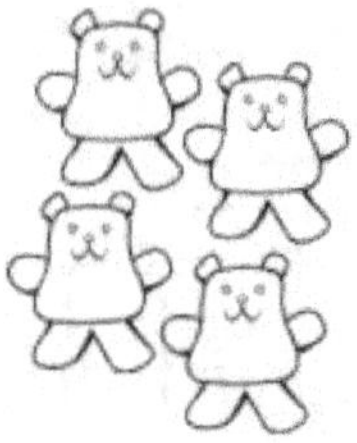

8

Addition Worksheets

4 + 1	3 + 10	3 + 7
8 + 10	6 + 10	7 + 10
10 + 2	10 + 3	4 + 9

Direction: Count the images. Write the number of images in the
boxes above each image and write the total number in the last box.

+   =

+   =

+   =

+   =

Name : _______________

Direction: Add the number of images in each box and
write the answer in the last box.

+  =

+  =

+  =

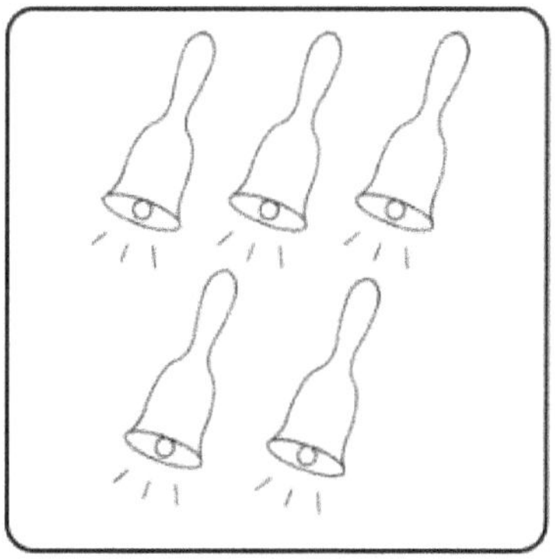

+  =

Addition Worksheets

5 + 1 [ ] Answer	5 + 5 [ ] Answer
2 + 5 [ ] Answer	5 + 5 [ ] Answer
1 + 3 [ ] Answer	3 + 3 [ ] Answer

Name : _______________________________

Direction: Add the images. Draw a line between the total number of images in each box and the number on the right.

9

5

9

8

Addition Worksheets

2
+ 1

4
+ 6

8
+ 6

10
+ 4

7
+ 10

7
+ 2

3
+ 10

4
+ 9

9
+ 6

Math Made Easy....

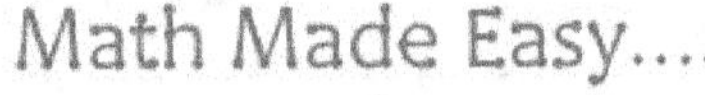

Name : _______________________________

Direction: Count the images. Write the number of images in the
boxes above each image and write the total number in the last box.

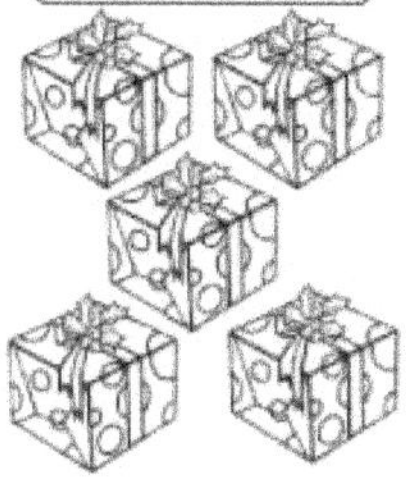

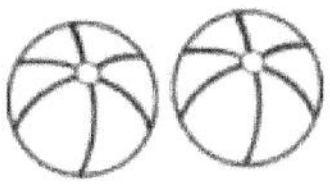

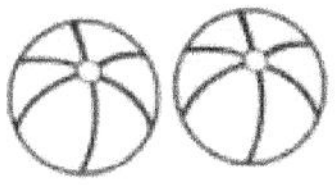

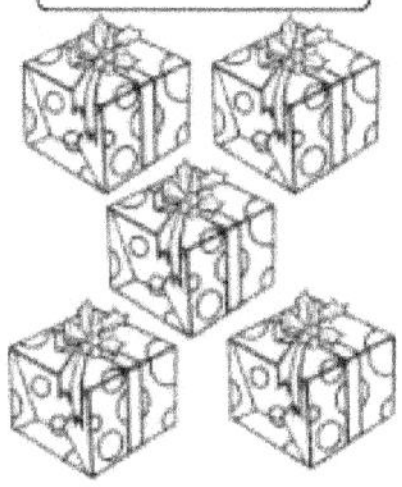

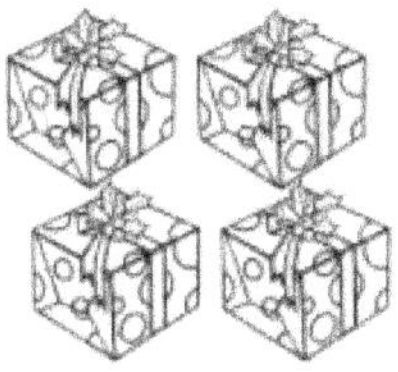

Name : _______________________________

Direction: Add the number of images in each box and
write the answer in the last box.

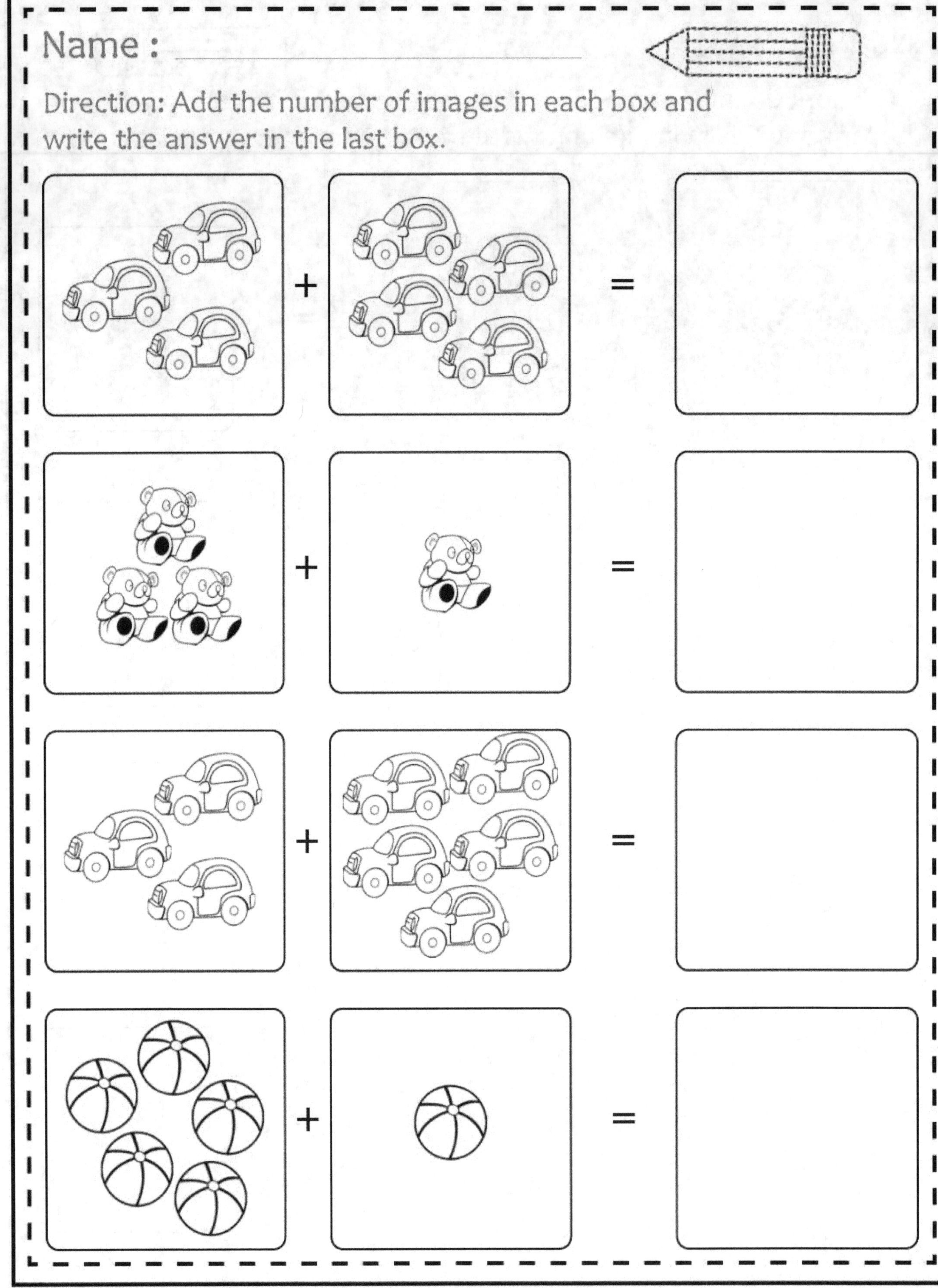

Addition Worksheets

5
+ 5

Answer

2
+ 5

Answer

2
+ 4

Answer

2
+ 1

Answer

5
+ 4

Answer

5
+ 5

Answer

Direction: Add the images. Draw a line between the total
number of images in each box and the number on the right.

6

4

5

6

Addition Worksheets

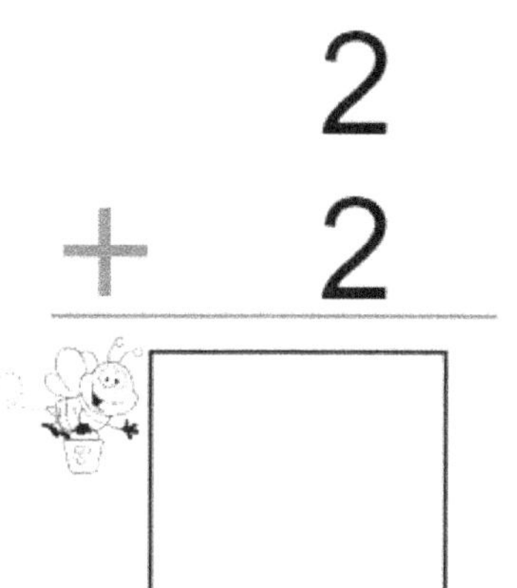

```
   2
+  2
______
```

```
   7
+ 10
______
```

```
   8
+  4
______
```

```
   5
+  3
______
```

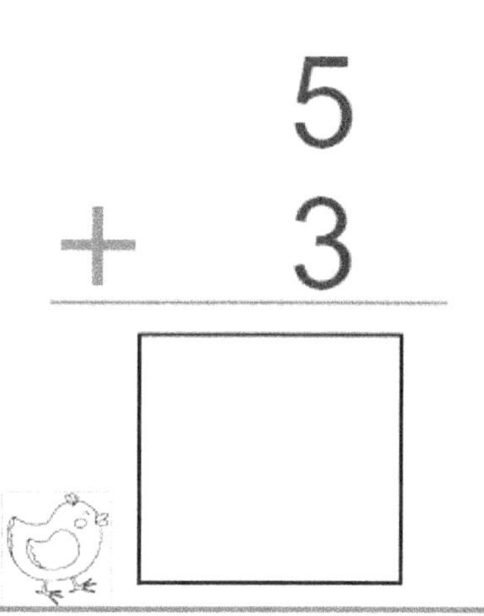

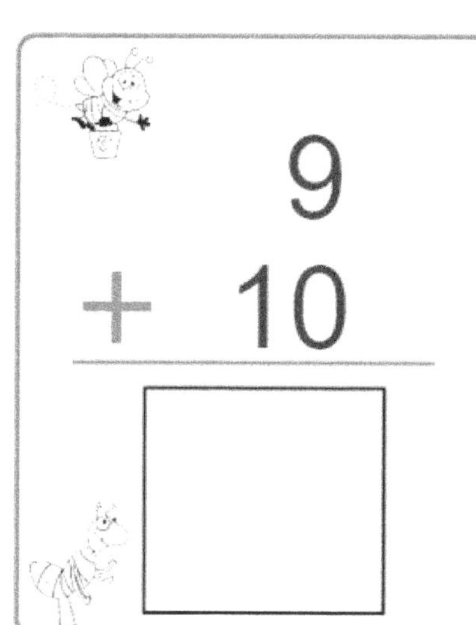

```
   9
+ 10
______
```

```
   2
+  5
______
```

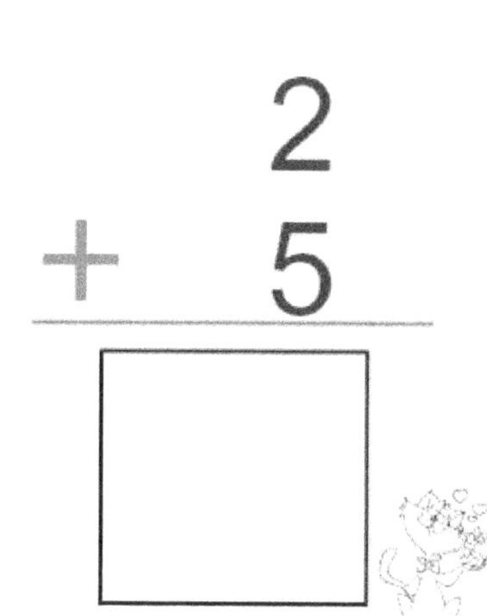

```
   9
+  5
______
```

```
   7
+  7
______
```

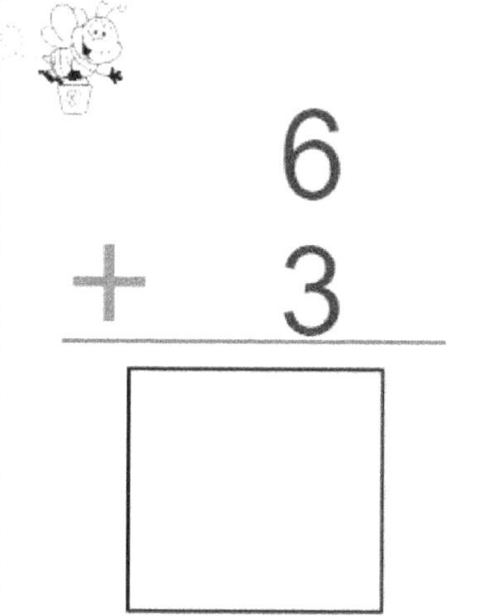

```
   6
+  3
______
```

Math Made Easy....

Name : _______________________________

Direction: Count the images. Write the number of images in the boxes above each image and write the total number in the last box.

+    =

+    =

+ 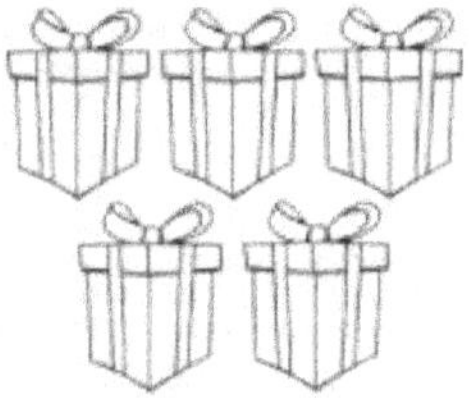   =

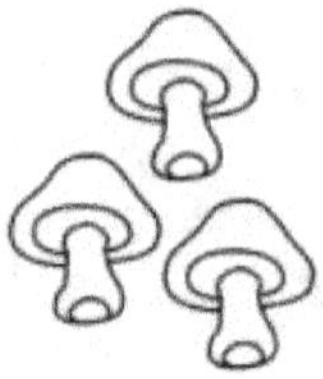

+    =

Name : __________

Direction: Add the number of images in each box and write the answer in the last box.

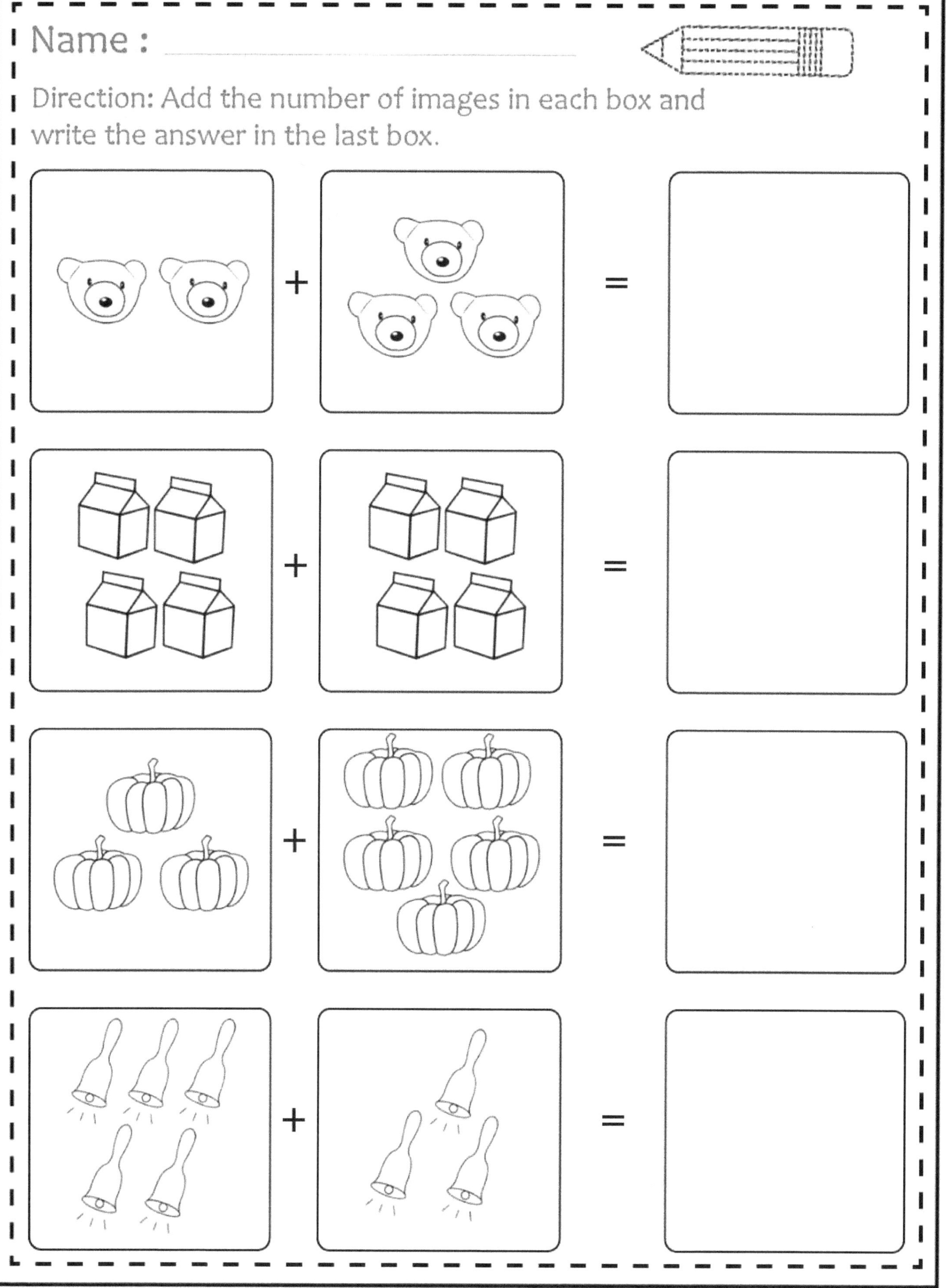

Addition Worksheets

Name : __________________________

2

+ 2

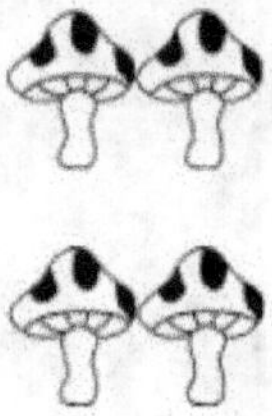

Answer

3

+ 5

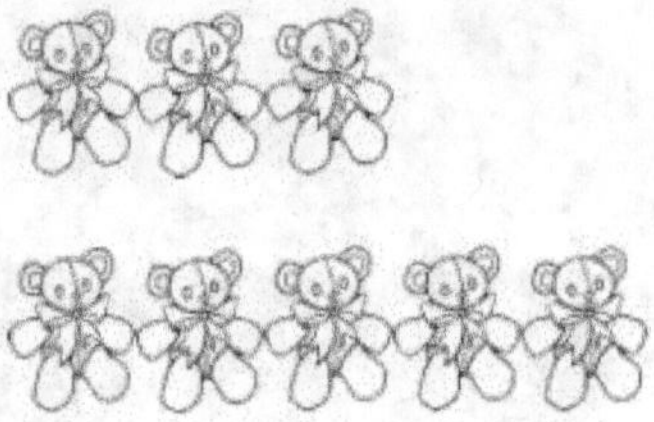

Answer

5

+ 5

Answer

4

+ 5

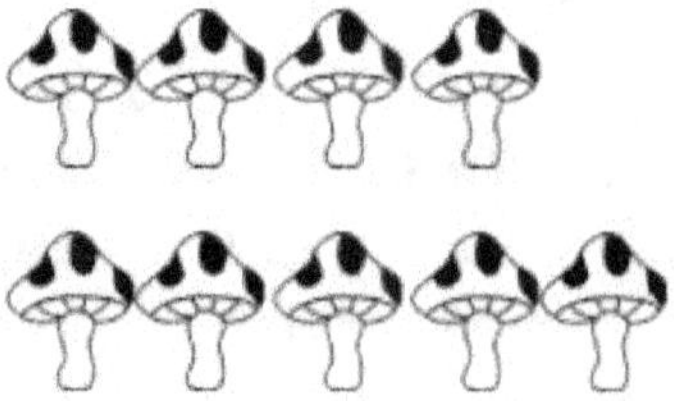

Answer

5

+ 5

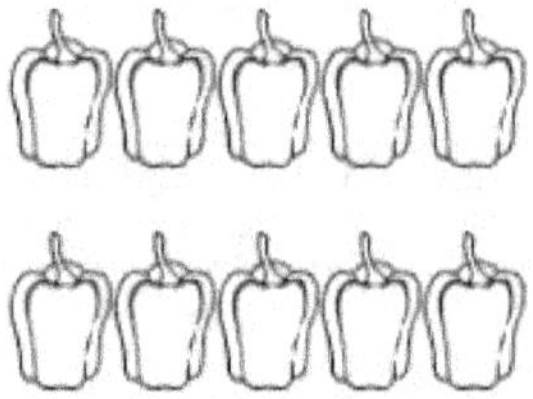

Answer

2

+ 1

Answer

Name : _______________________

Direction: Add the images. Draw a line between the total
number of images in each box and the number on the right.

7

7

5

3

Addition Worksheets

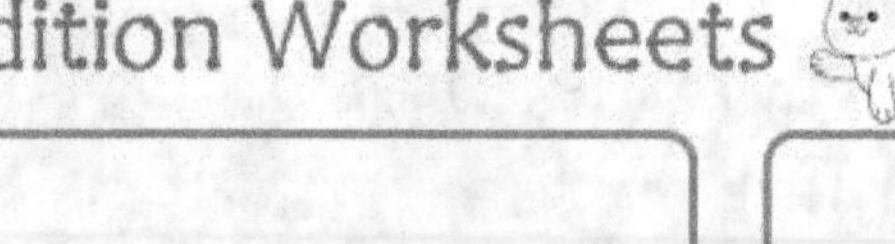

2
+ 3

8
+ 5

5
+ 7
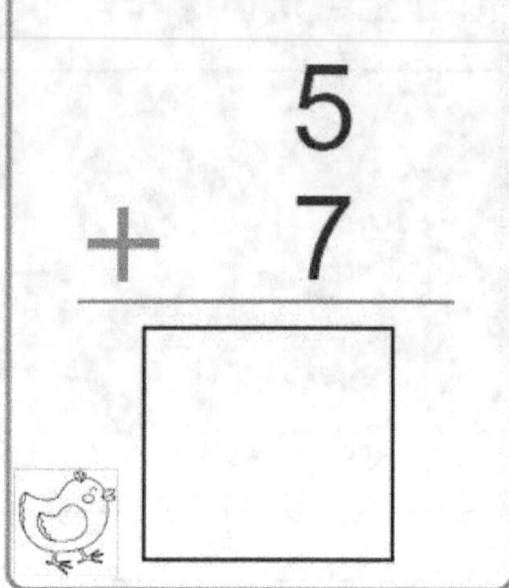

4
+ 2
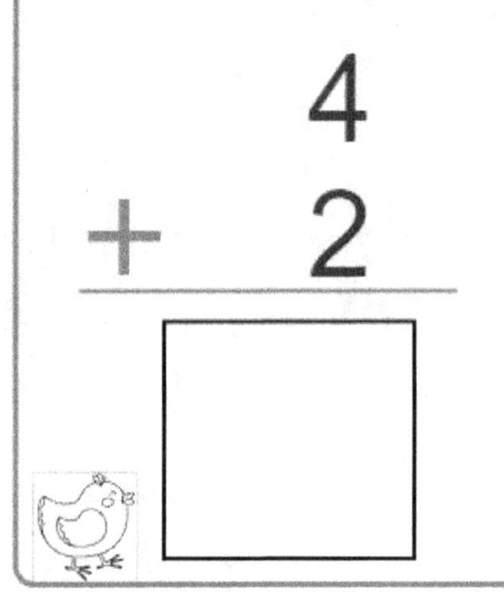

7
+ 7
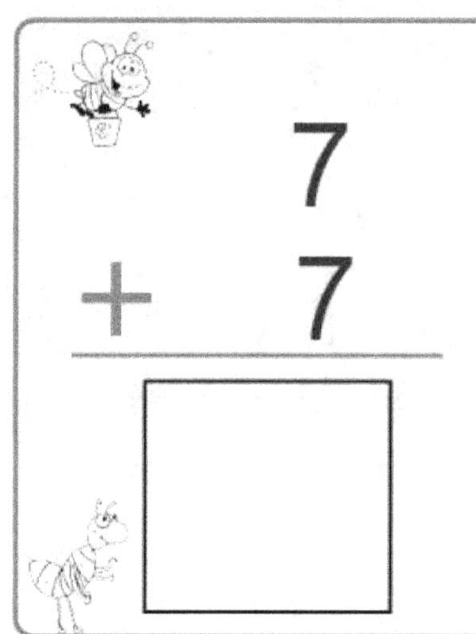

9
+ 7
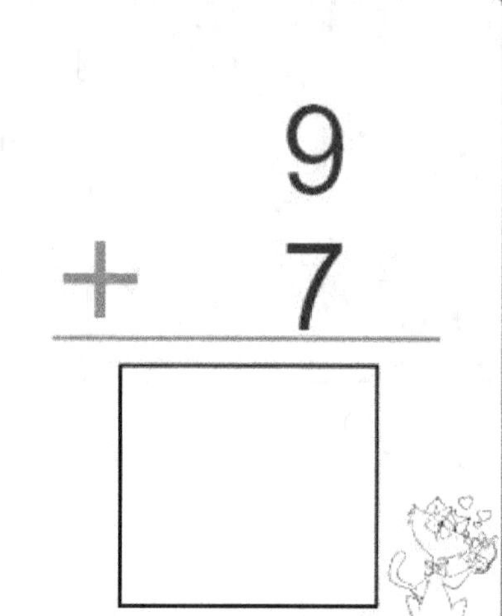

8
+ 4

2
+ 9

4
+ 10
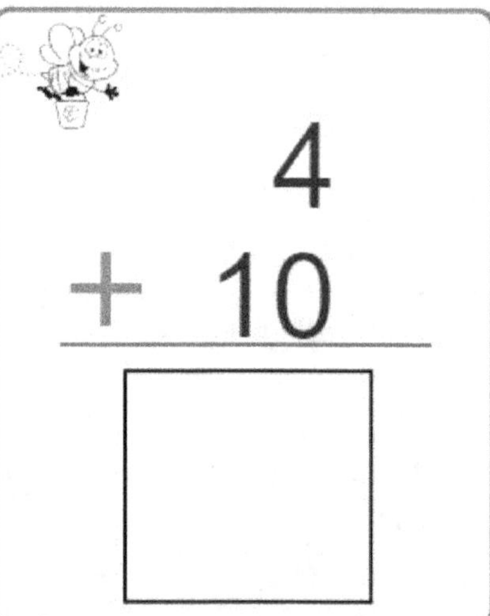

 Math Made Easy.... 

Name : ___________________________

Direction: Count the images. Write the number of images in the
boxes above each image and write the total number in the last box.

Name :
Direction: Add the number of images in each box and write the answer in the last box.

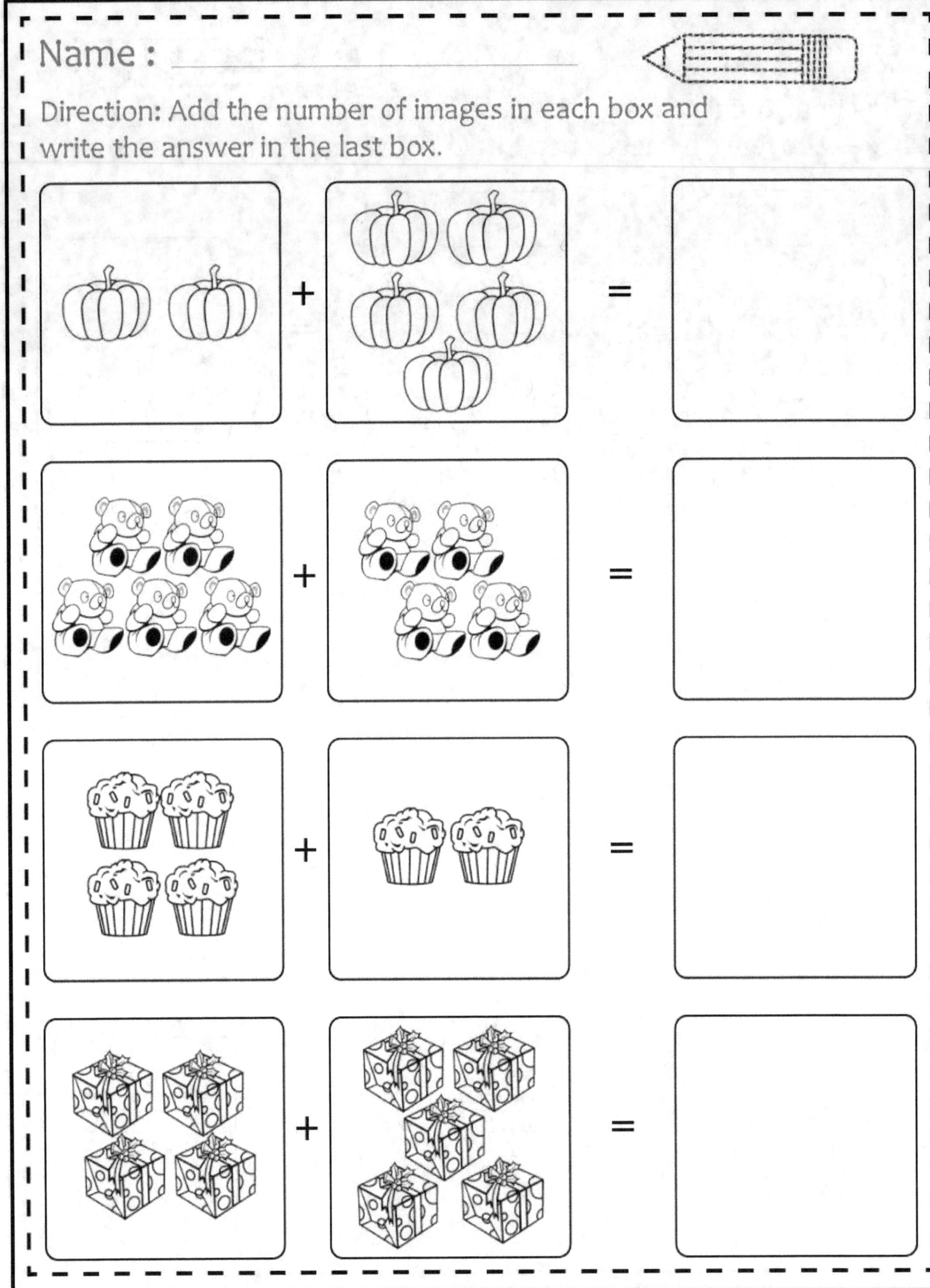

Name : _______________

Addition Worksheets

3 + 2

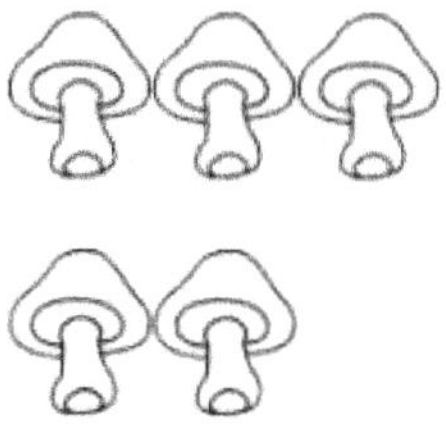

Answer

3 + 4

Answer

4 + 4

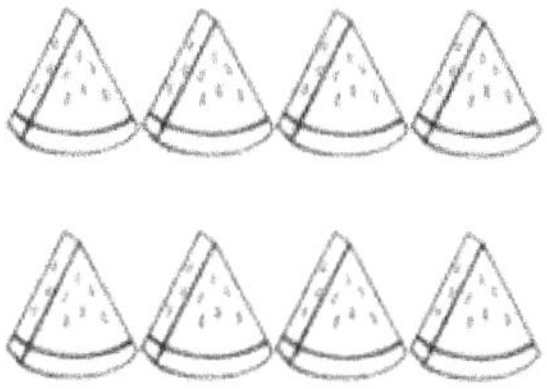

Answer

1 + 5

Answer

1 + 5

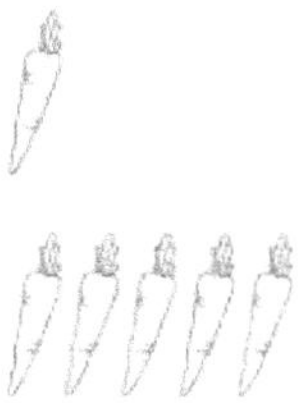

Answer

2 + 1

Answer

Name : _______________________________

Direction: Add the images. Draw a line between the total
number of images in each box and the number on the right.

8

6

9

3

Addition Worksheets

3 + 5	2 + 5	6 + 10
9 + 9	9 + 1	1 + 8
2 + 7	3 + 2	2 + 8

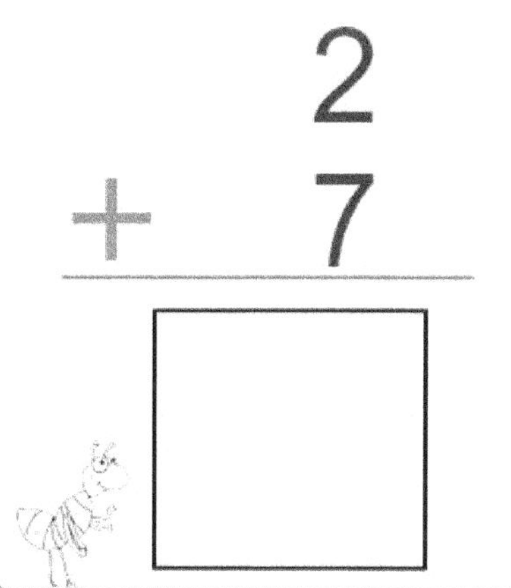
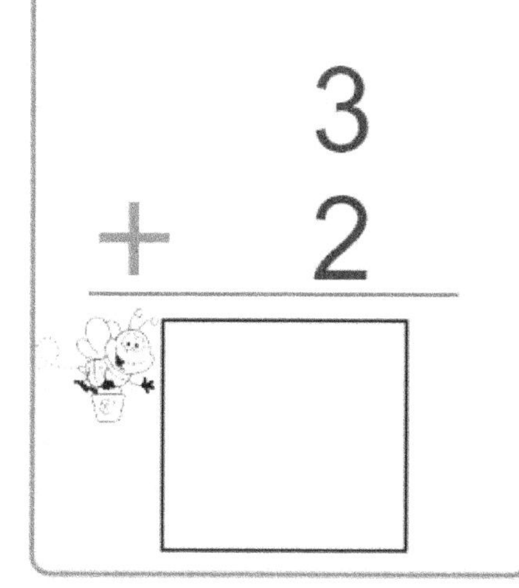
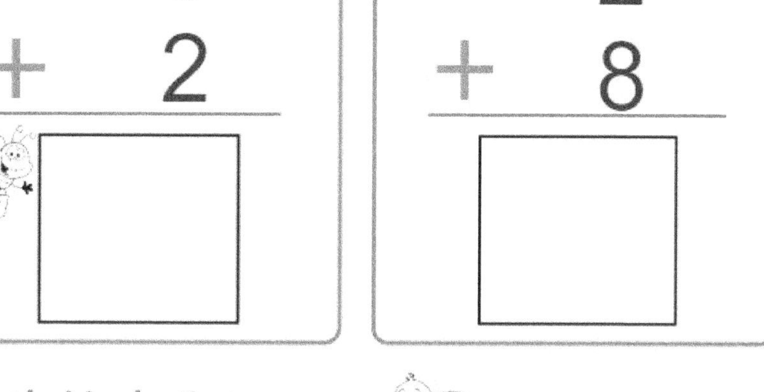

Math Made Easy....

Name : _______________________

Direction: Count the images. Write the number of images in the boxes above each image and write the total number in the last box.

+   =

+   =

+   =

+   =

Name : _______________________

Direction: Add the number of images in each box and
write the answer in the last box.

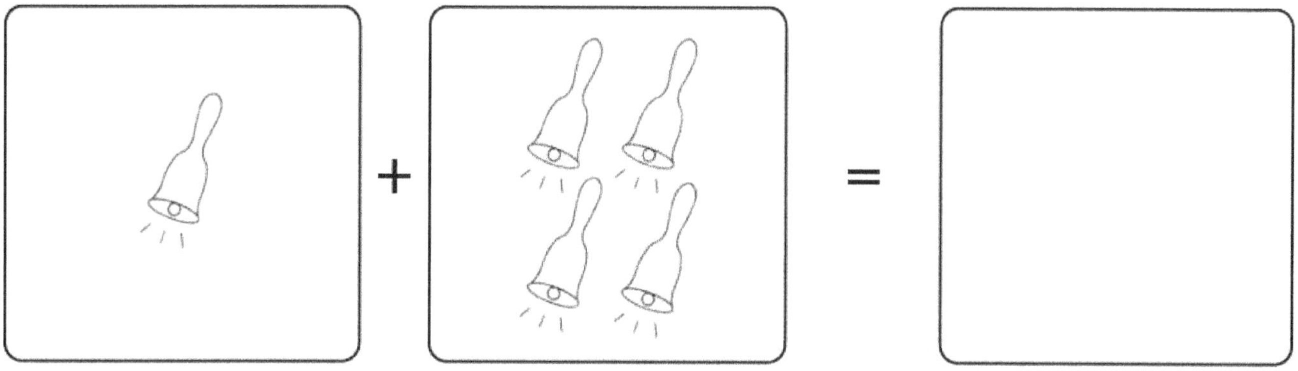

Addition Worksheets

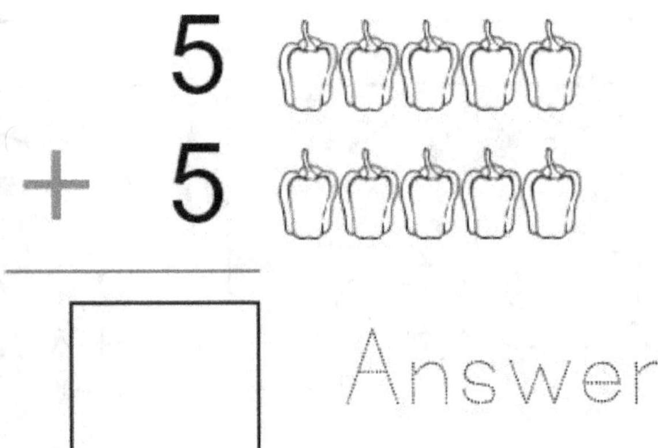

Direction: Add the images. Draw a line between the total number of images in each box and the number on the right.

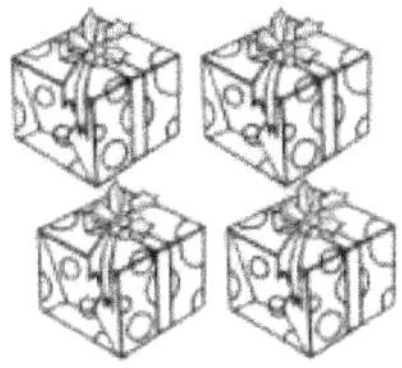

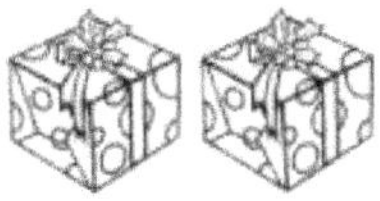

3

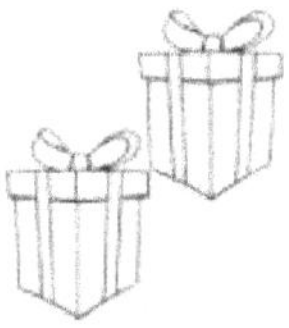

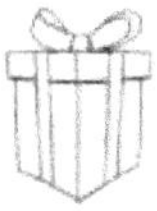

8

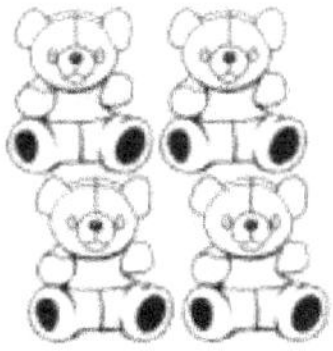

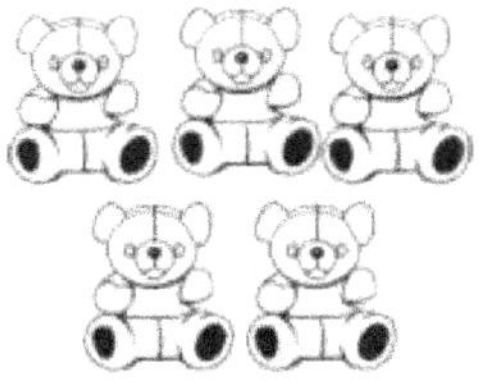

9

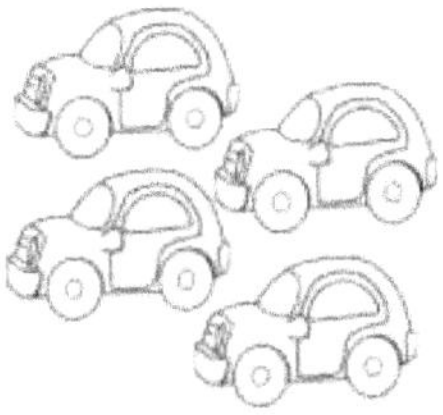

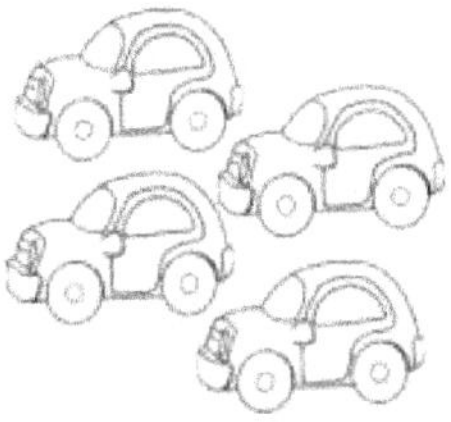

6

Addition Worksheets

10 + 5 = ☐	1 + 8 = ☐	9 + 3 = ☐
10 + 1 = ☐	2 + 3 = ☐	2 + 9 = ☐
7 + 1 = ☐	9 + 1 = ☐	3 + 5 = ☐

Direction: Count the images. Write the number of images in the boxes above each image and write the total number in the last box.

+ =

+ =

+ =

+ =

Direction: Add the number of images in each box and write the answer in the last box.

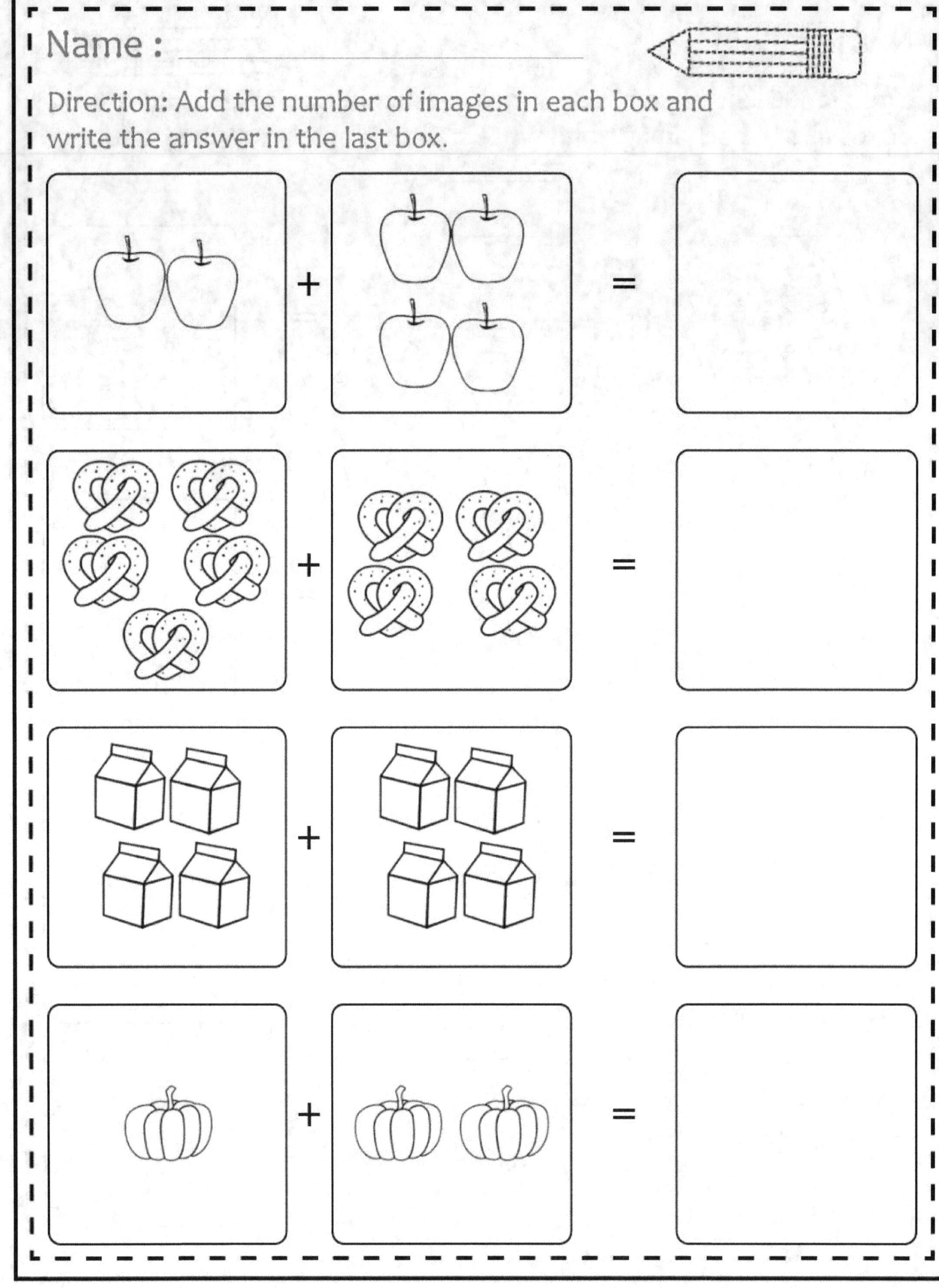

Name : _______________

Addition Worksheets

3
+ 5

Answer

1
+ 5

Answer

5
+ 5

Answer

4
+ 3

Answer

1
+ 5

Answer

4
+ 4

Answer

Name :

Direction: Add the images. Draw a line between the total number of images in each box and the number on the right.

+ • 3

+ • 8

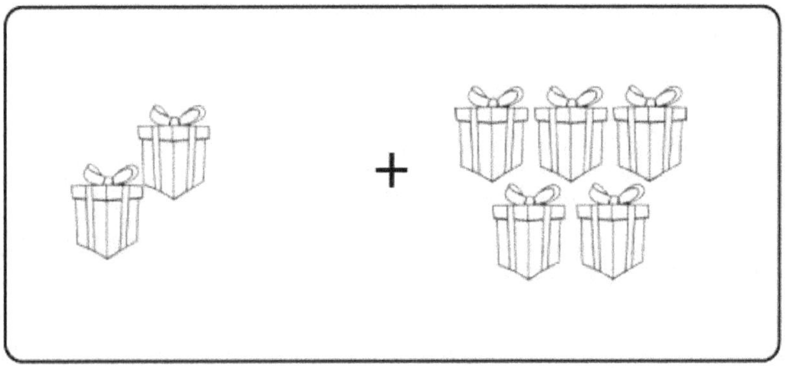

+ • 7

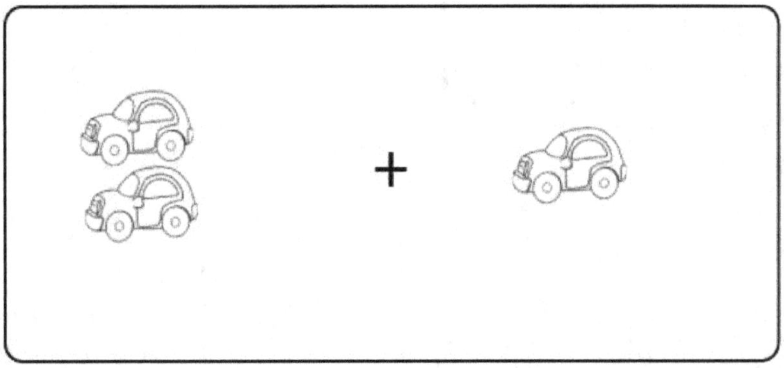

+ • 7

Addition Worksheets

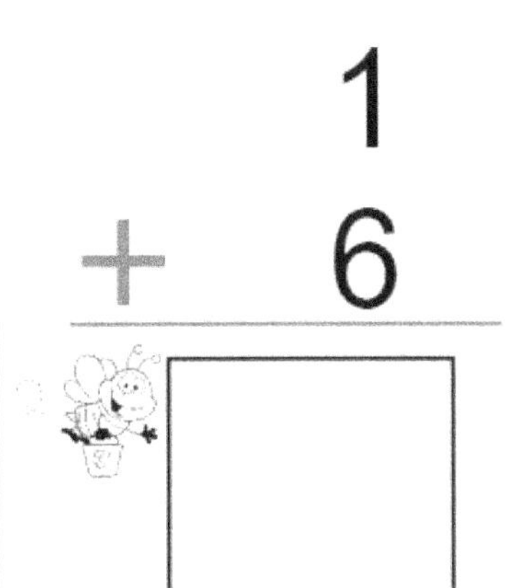

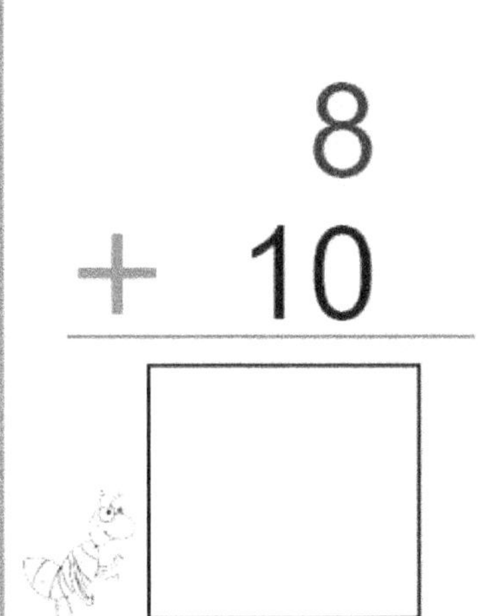

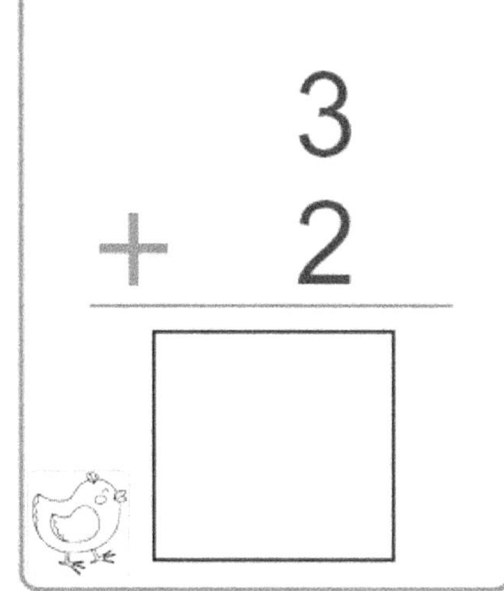

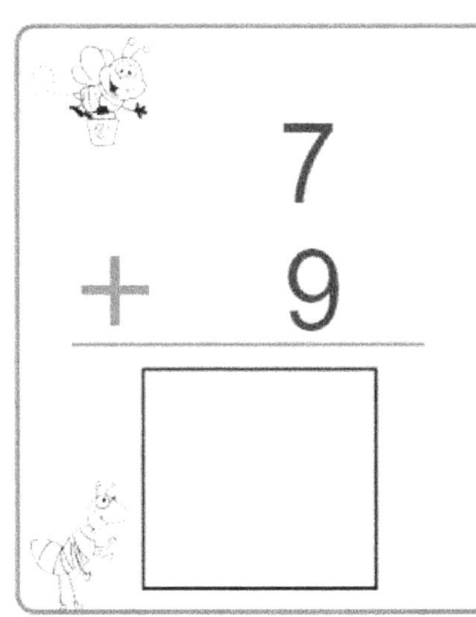

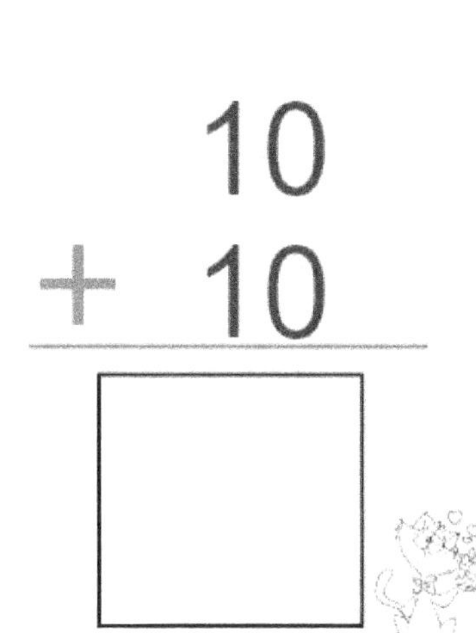

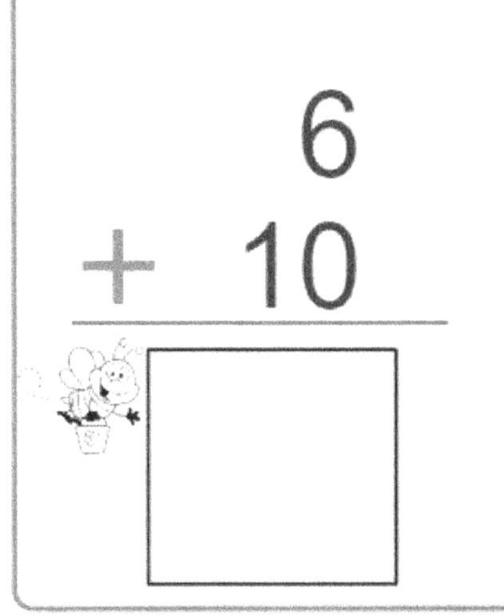

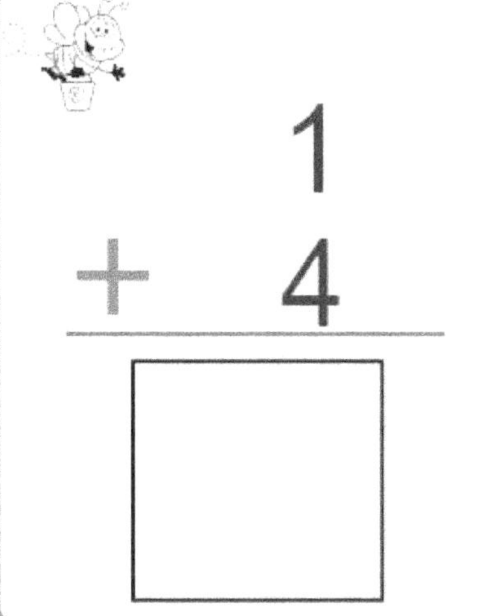

Math Made Easy....

Name : _______________________

Direction: Count the images. Write the number of images in the boxes above each image and write the total number in the last box.

+ =

+ =

+ =

+ =

Name :
Direction: Add the number of images in each box and
write the answer in the last box.
+
=
+
=
+
=
+
=

Name : _______________________

Addition Worksheets

$$4 + 3 =$$ Answer

$$3 + 5 =$$ Answer

$$2 + 4 =$$ Answer

$$2 + 5 =$$ Answer

$$4 + 2 =$$ Answer

$$5 + 1 =$$ Answer

Direction: Add the images. Draw a line between the total number of images in each box and the number on the right.

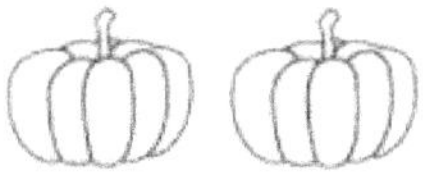

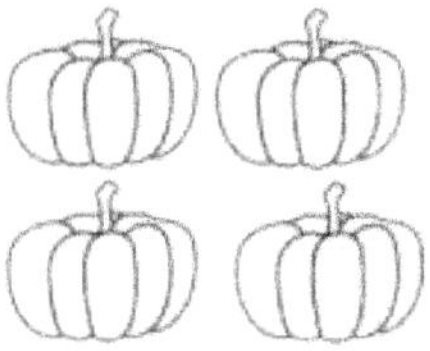

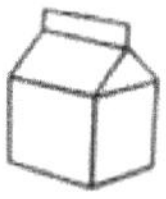

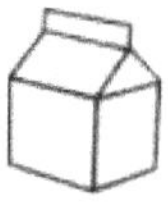

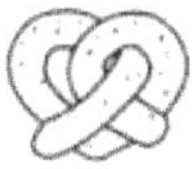

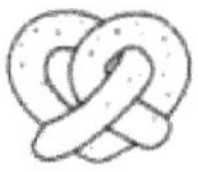

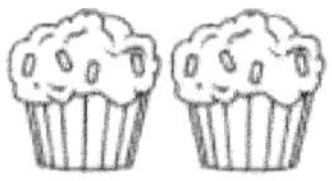

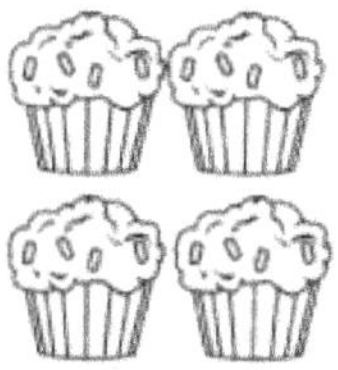

6 + 5	5 + 4	4 + 9
5 + 3	10 + 7	6 + 5
4 + 6	7 + 6	7 + 6

 Math Made Easy.... 

Name : _______________________

Direction: Count the images. Write the number of images in the boxes above each image and write the total number in the last box.

+   =

+   =

+   =

+   =

Name :
Direction: Add the number of images in each box and
write the answer in the last box.

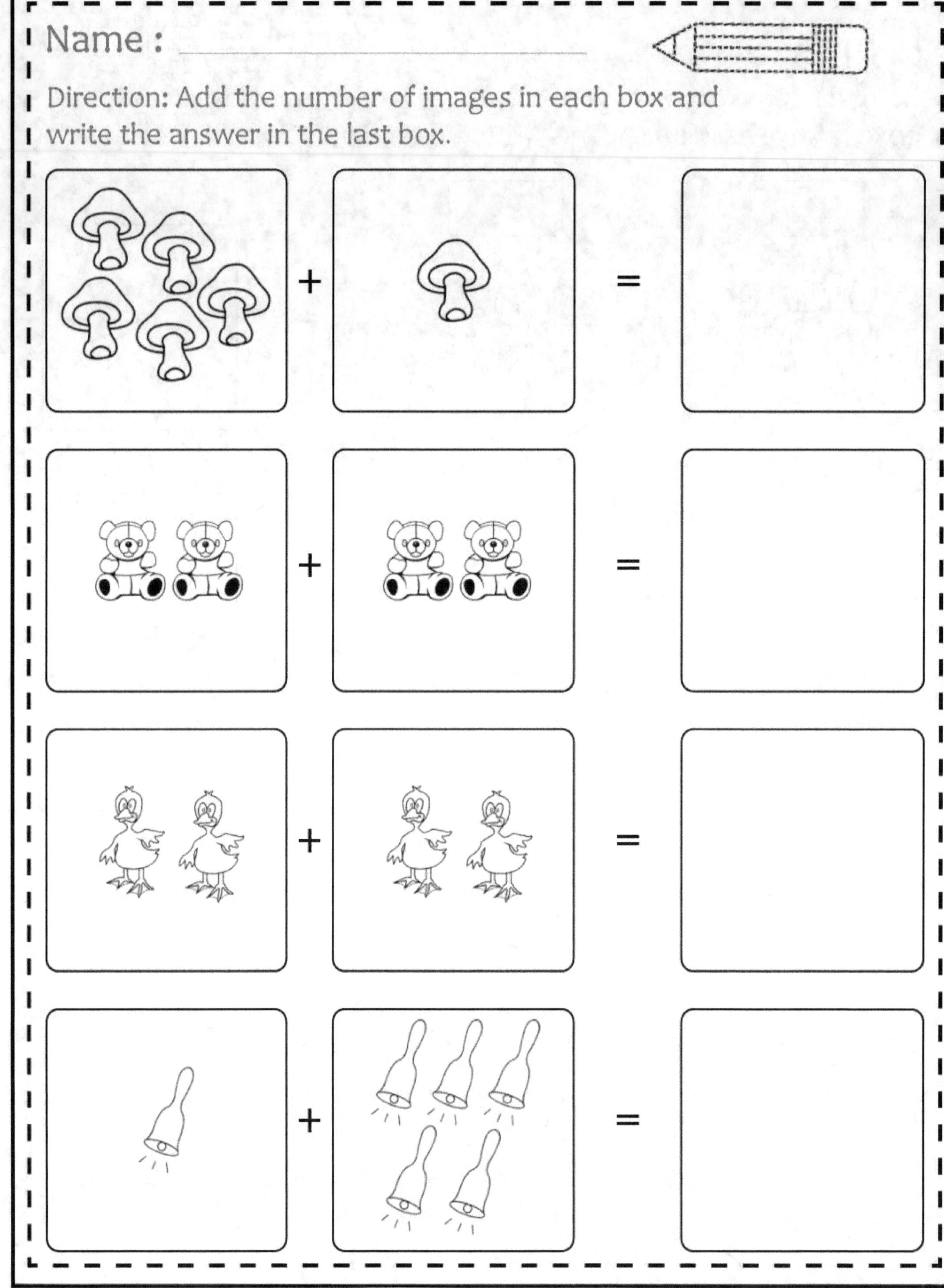

Addition Worksheets

$1 + 4 =$ Answer

$4 + 3 =$ Answer

$5 + 5 =$ Answer

$2 + 5 =$ Answer

$3 + 2 =$ Answer

$5 + 1 =$ Answer

Name : _______________________

Direction: Add the images. Draw a line between the total
number of images in each box and the number on the right.

6

6

7

6

Addition Worksheets

	7
+	2

	4
+	9

	3
+	9

	4
+	2

	6
+	6

	2
+	6

	9
+	8

	2
+	6

	8
+	4

Math Made Easy....

Name : ___________________________

Direction: Count the images. Write the number of images in the
boxes above each image and write the total number in the last box.

+   =

+   =

+   =

+   =

Name : _______________

Direction: Add the number of images in each box and
write the answer in the last box.

2
+ 5

Answer

5
+ 2

Answer

5
+ 1

Answer

5
+ 4

Answer

5
+ 1

Answer

5
+ 3

Answer

Name : _______________________

Direction: Add the images. Draw a line between the total
number of images in each box and the number on the right.

3

4

4

7

Addition Worksheets

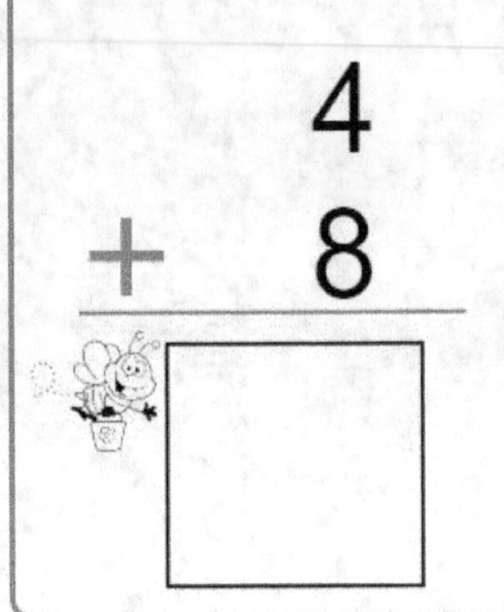

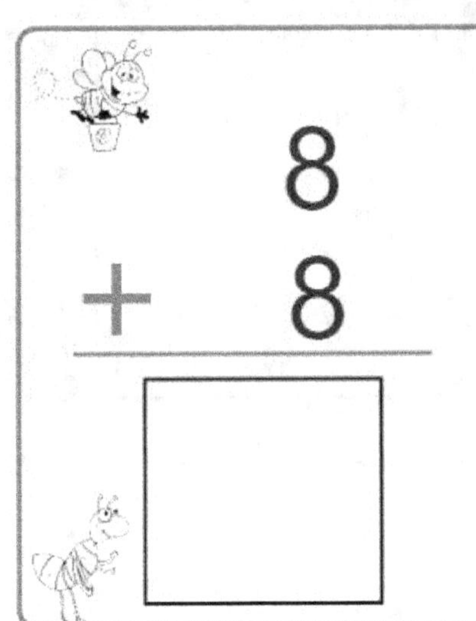

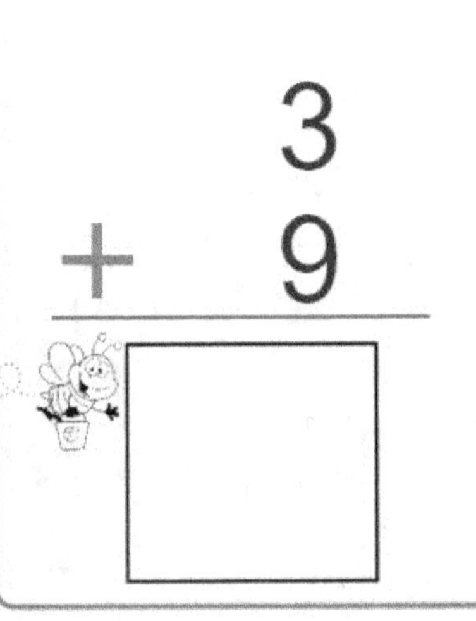

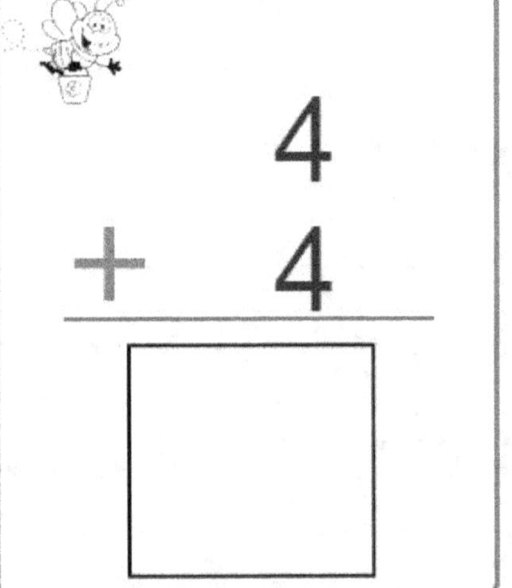

 Math Made Easy.... 

Name : _______________________________

Direction: Count the images. Write the number of images in the
boxes above each image and write the total number in the last box.

+  =

+  =

+  =

+  =

Name : _______________________________

Direction: Add the number of images in each box and
write the answer in the last box.

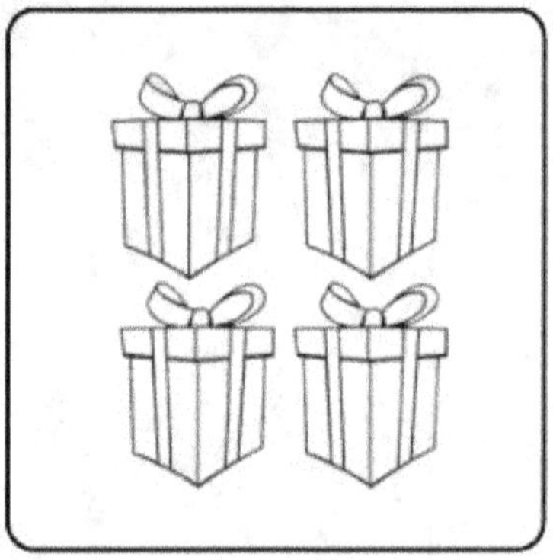

Name : ________________

Addition Worksheets

5
+ 2

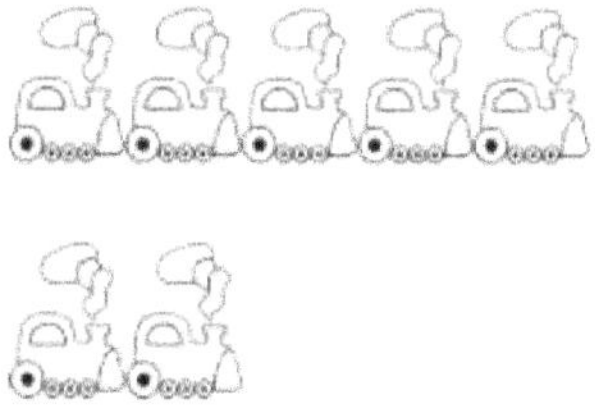

☐ Answer

2
+ 3

☐ Answer

1
+ 1

☐ Answer

4
+ 1

☐ Answer

5
+ 5

☐ Answer

3
+ 4

☐ Answer

Name : ________________________

Direction: Add the images. Draw a line between the total
number of images in each box and the number on the right.

5

7

4

9

Addition Worksheets

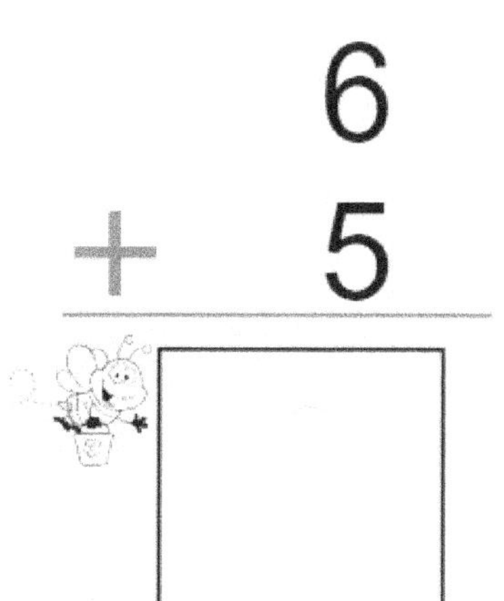

6
+ 5

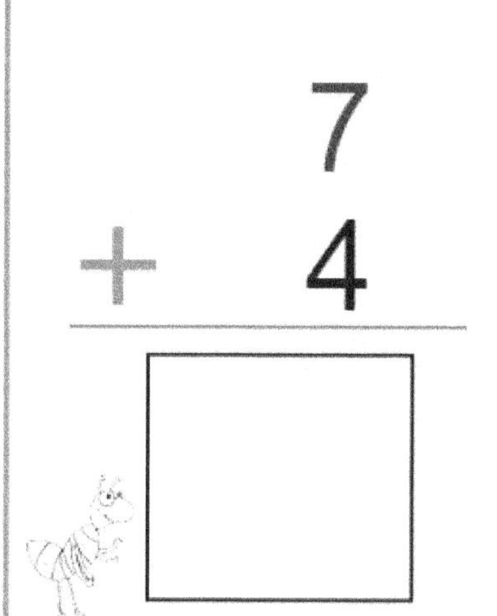

7
+ 4

5
+ 1

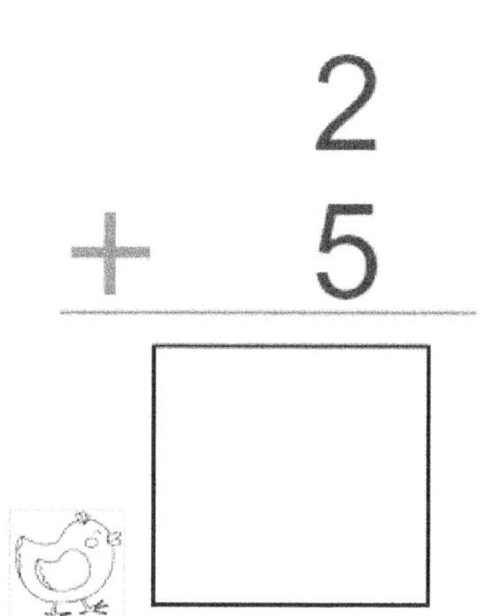

2
+ 5

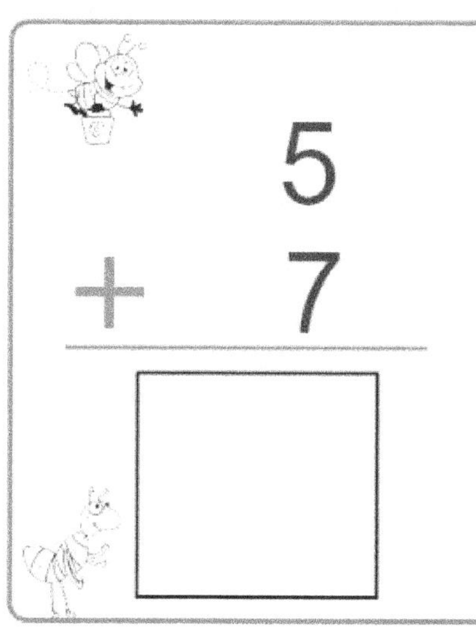

5
+ 7

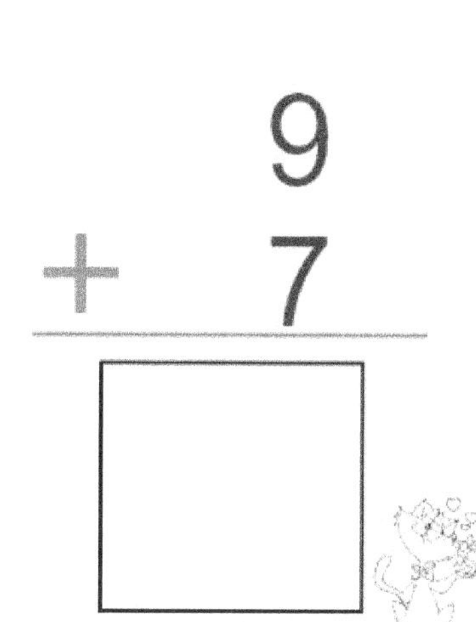

9
+ 7

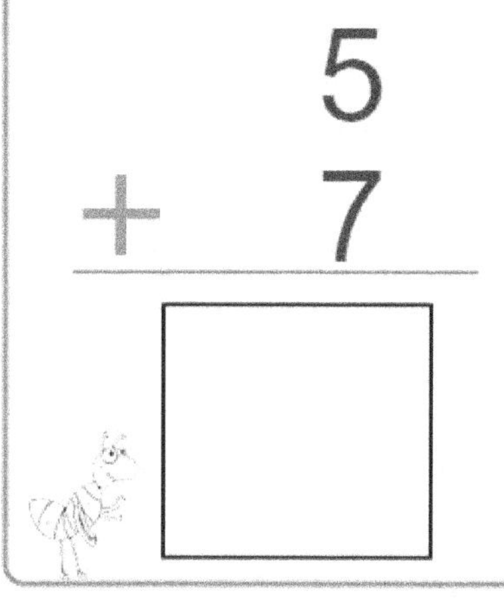

5
+ 7

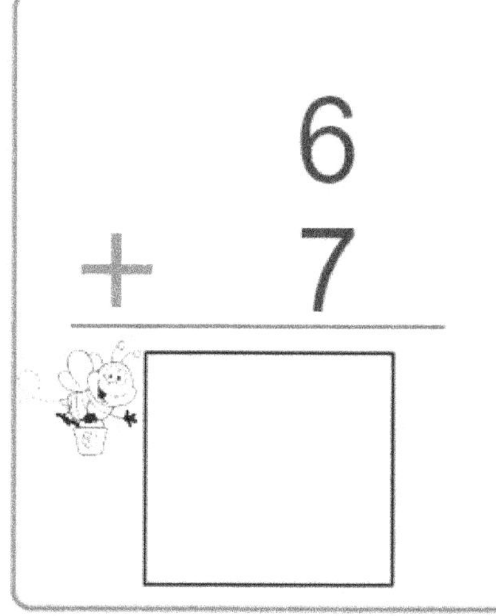

6
+ 7

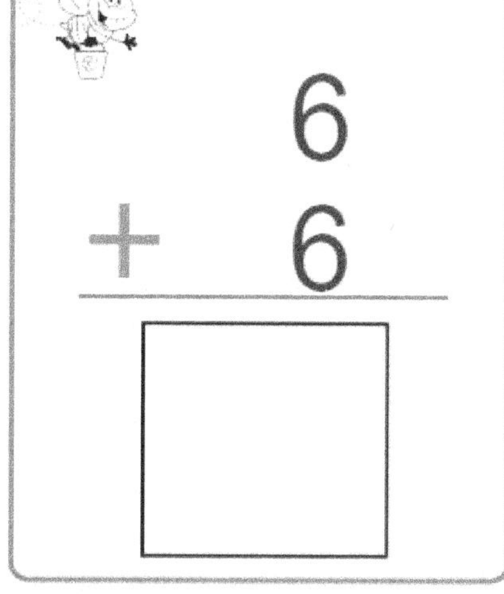

6
+ 6

Math Made Easy....

Name : __________________________

Direction: Count the images. Write the number of images in the boxes above each image and write the total number in the last box.

+ =

+ =

+ =

+ =

Name : _______________

Direction: Add the number of images in each box and write the answer in the last box.

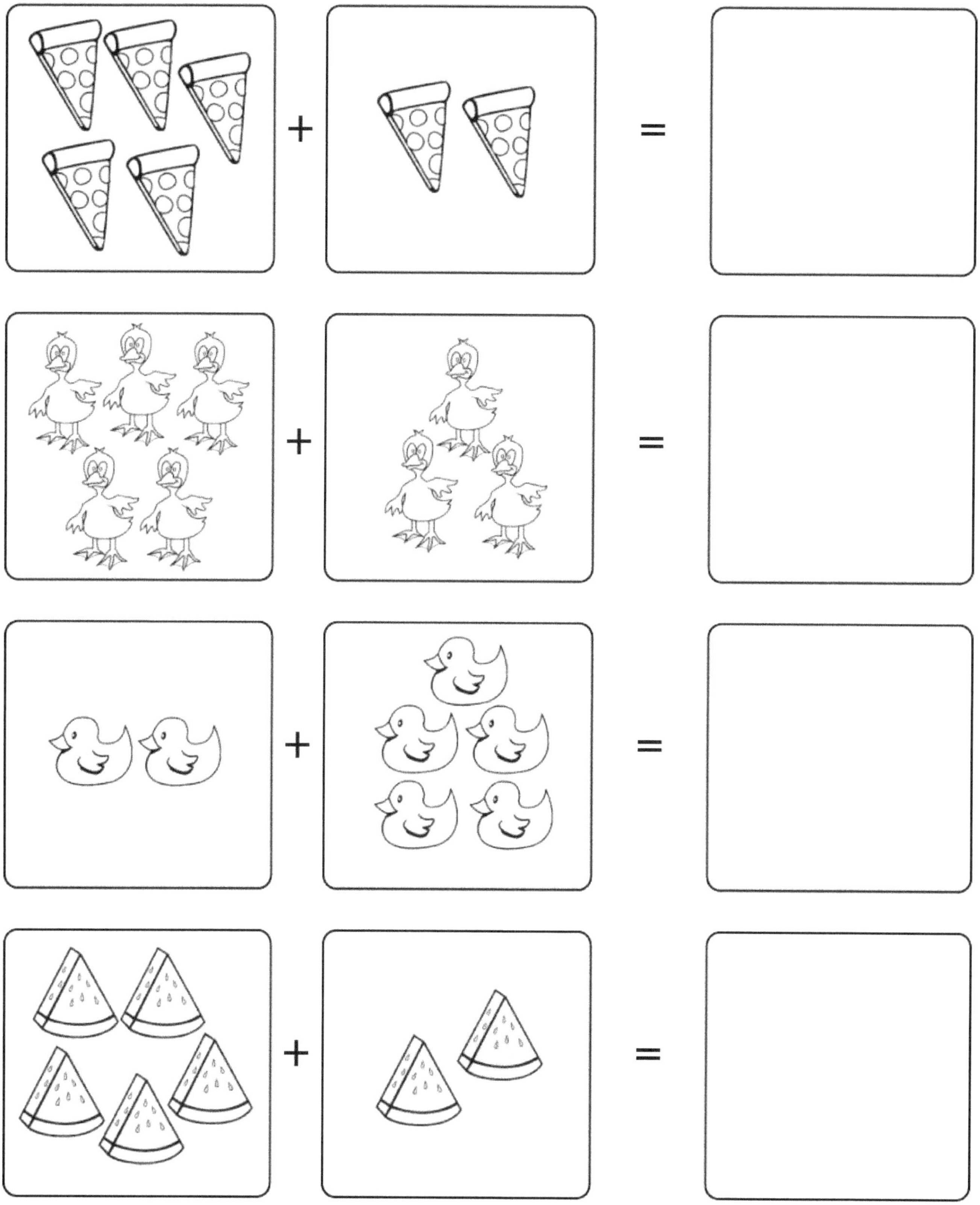

+   =

+   =

+   =

+   =

Addition Worksheets

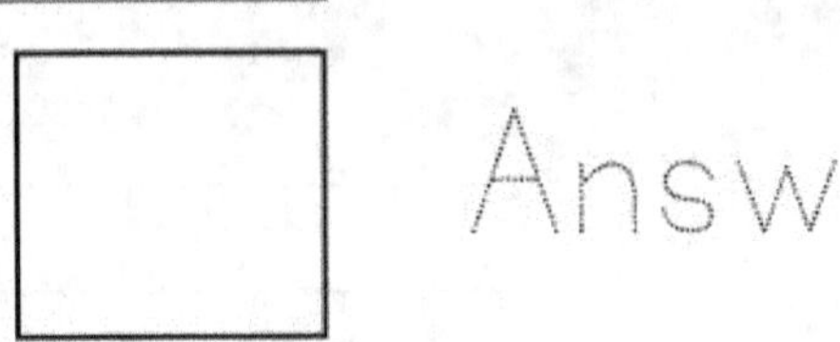

5
+ 2
Answer

3
+ 5
Answer

4
+ 5
Answer

5
+ 5
Answer

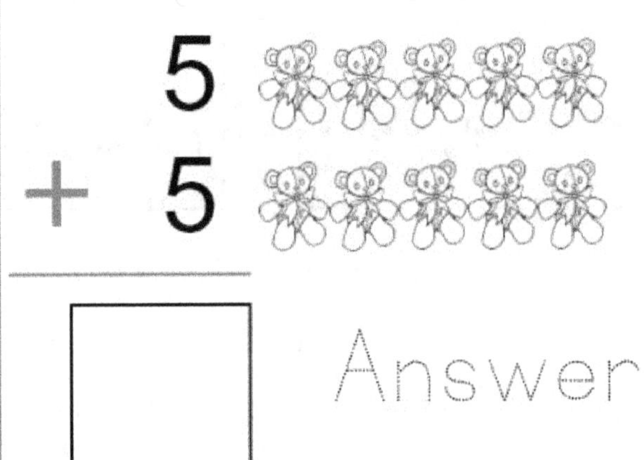

4
+ 4
Answer

4
+ 5
Answer

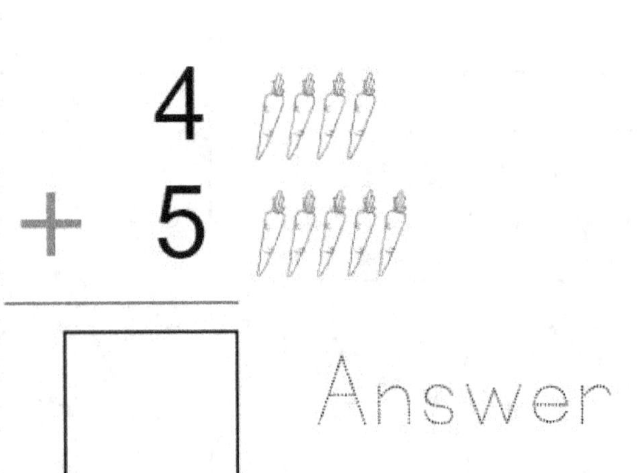

Name : _______________________

Direction: Add the images. Draw a line between the total
number of images in each box and the number on the right.

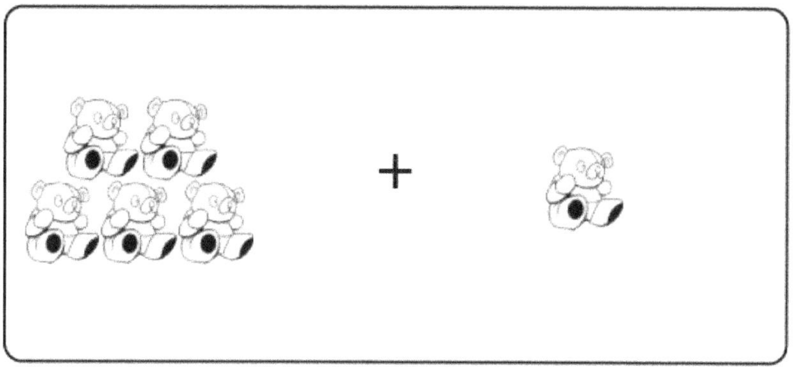

5

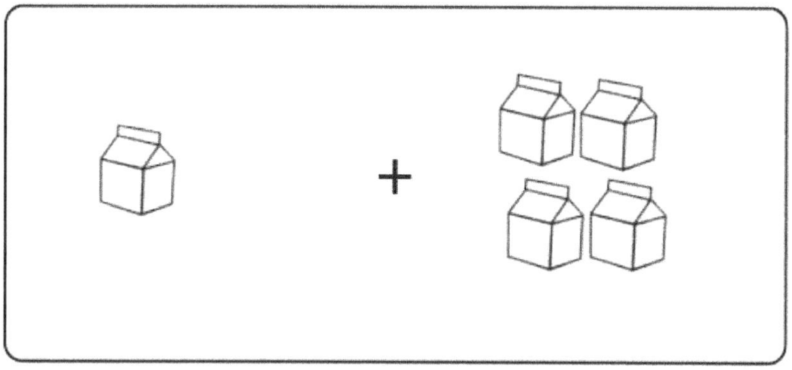

4

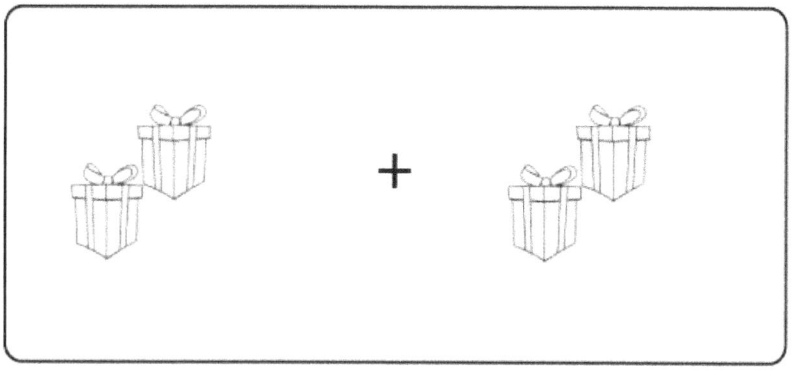

8

6

Name : _______________________

Addition Worksheets

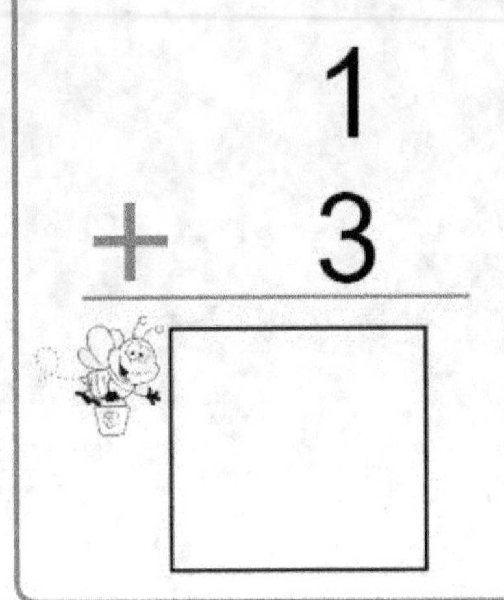

$$1 + 3 =$$

$$3 + 8 =$$

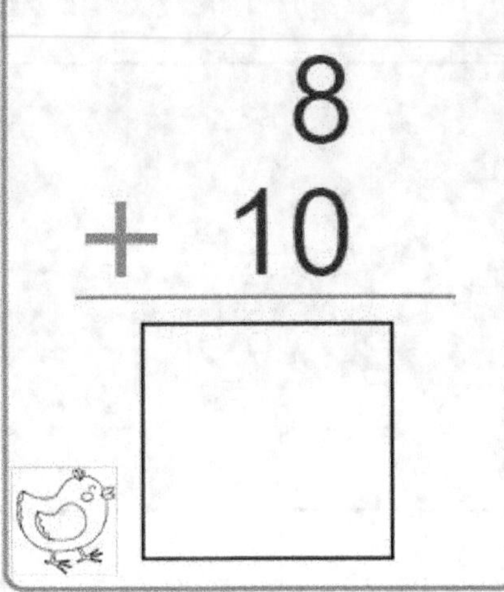

$$8 + 10 =$$

$$4 + 5 =$$

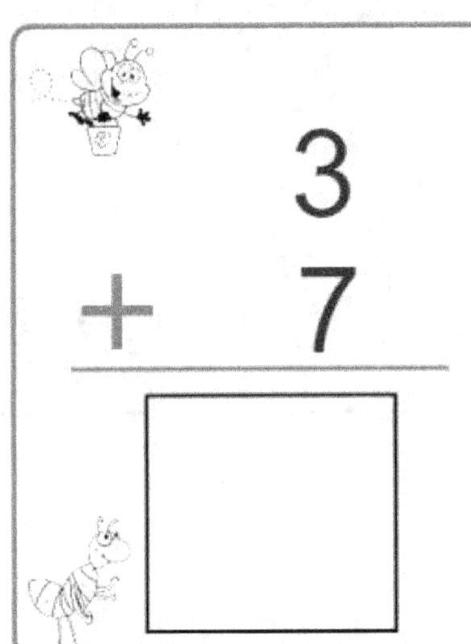

$$3 + 7 =$$

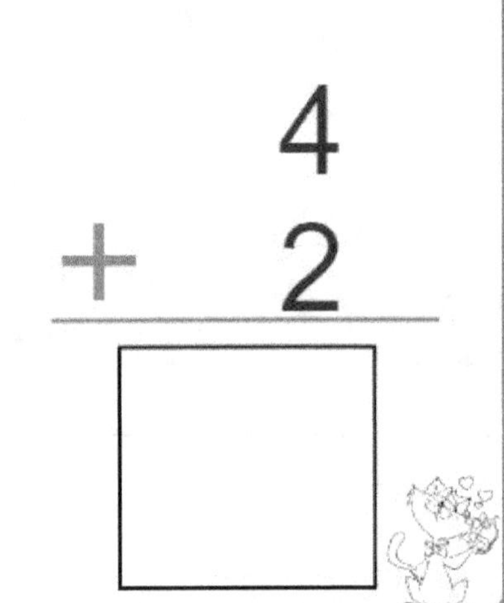

$$4 + 2 =$$

$$3 + 8 =$$

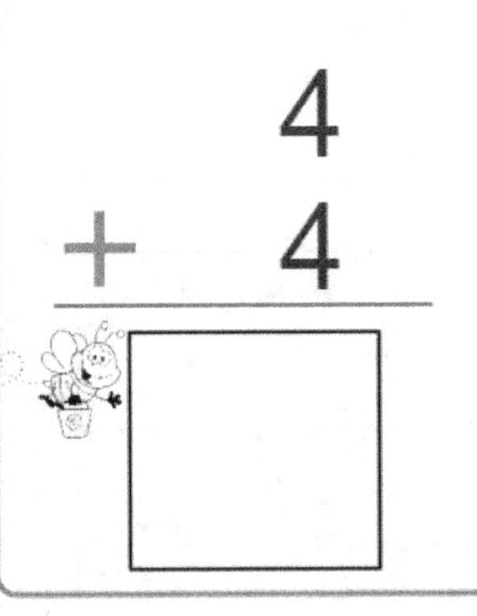

$$4 + 4 =$$

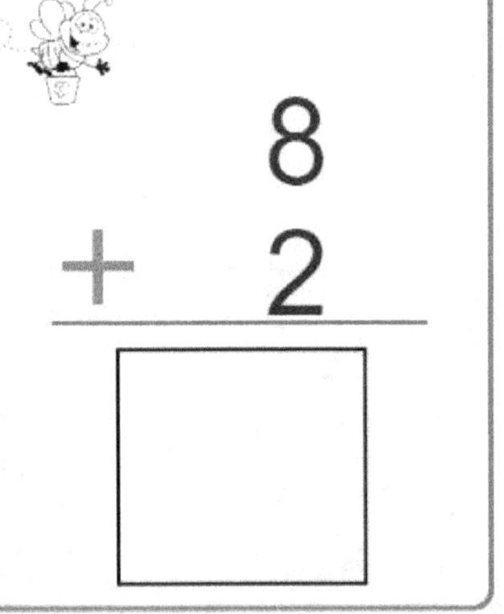

$$8 + 2 =$$

Math Made Easy....

Name : _______________________________

Direction: Count the images. Write the number of images in the
boxes above each image and write the total number in the last box.

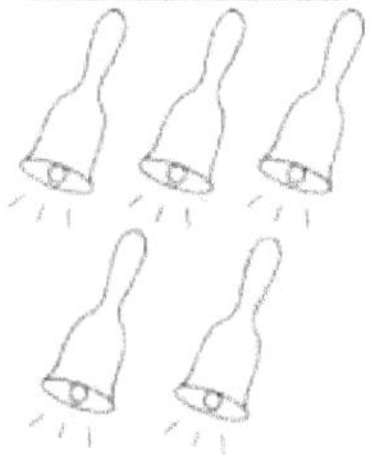

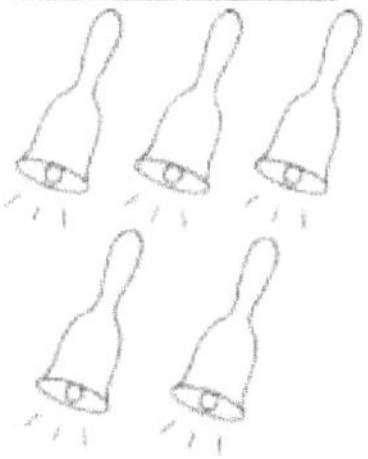

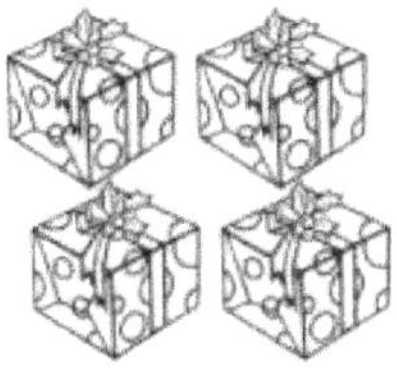

Direction: Add the number of images in each box and write the answer in the last box.

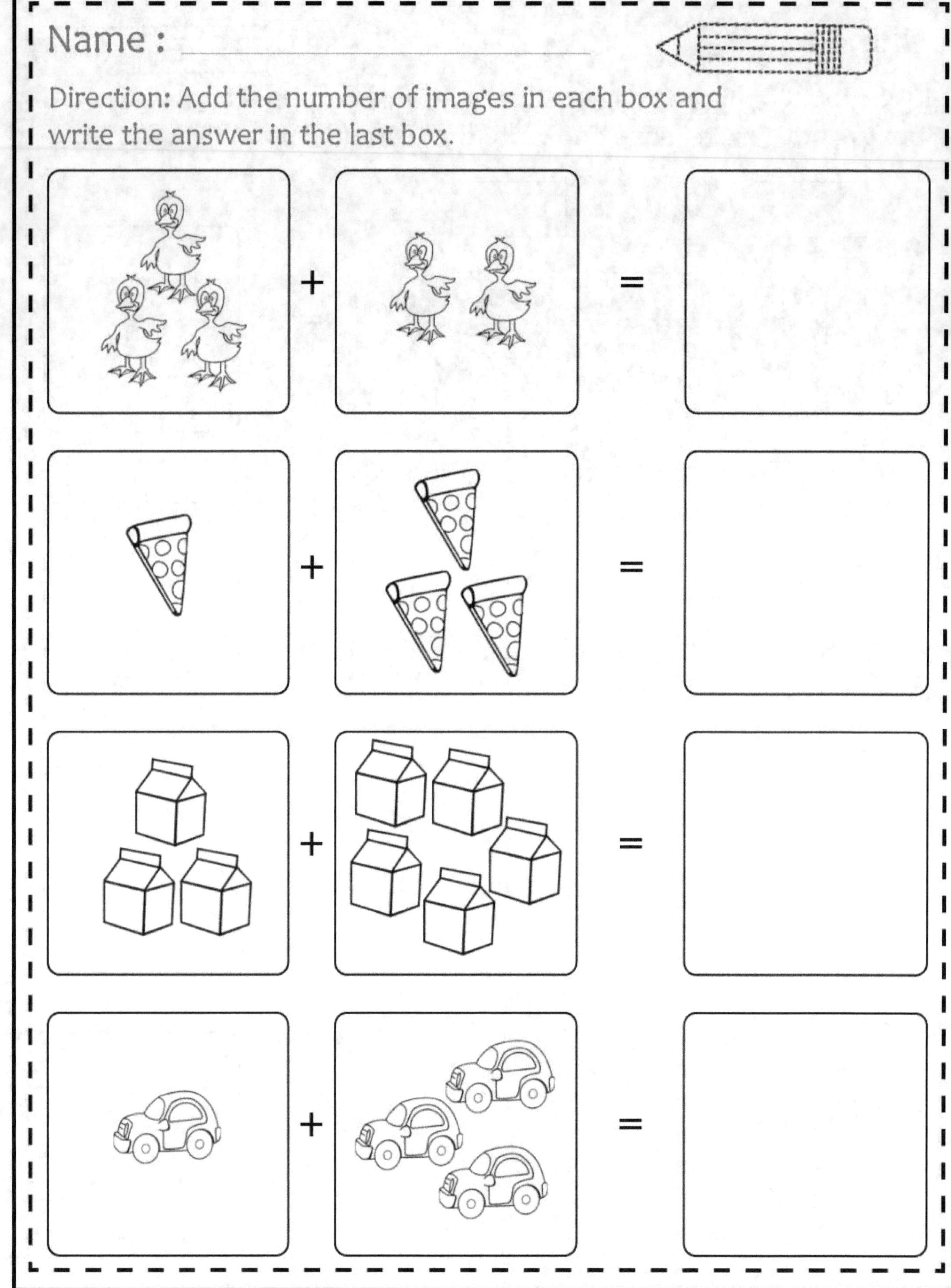

Name : _______________

Addition Worksheets

5
+ 5

Answer

5
+ 3

Answer

2
+ 5

Answer

3
+ 5

Answer

4
+ 5

Answer

5
+ 5

Answer

Name : __________________________________

Direction: Add the images. Draw a line between the total
number of images in each box and the number on the right.

4

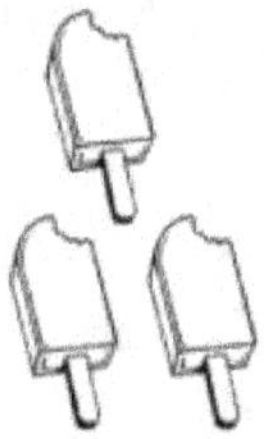

6

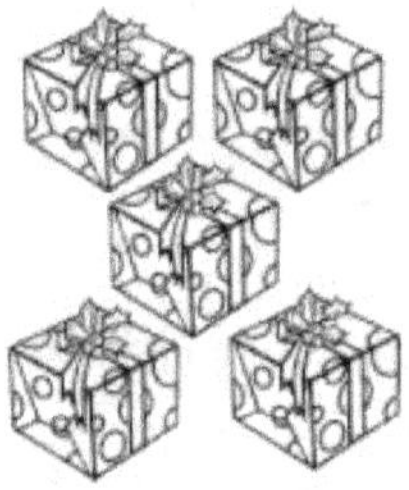

7

6

Addition Worksheets

8 + 1	7 + 5	8 + 3
8 + 9	6 + 6	1 + 9
8 + 1	3 + 1	7 + 8

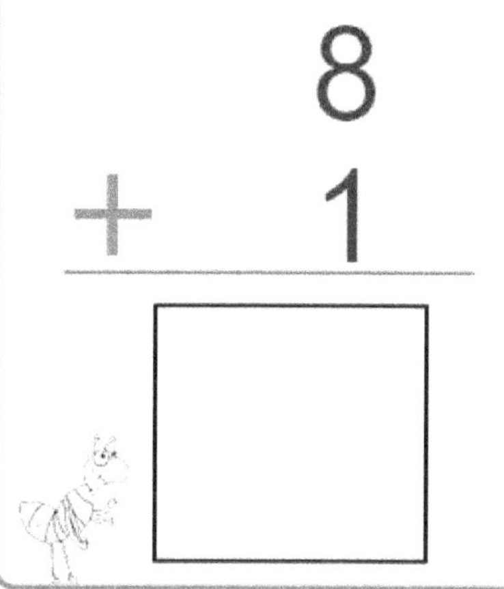
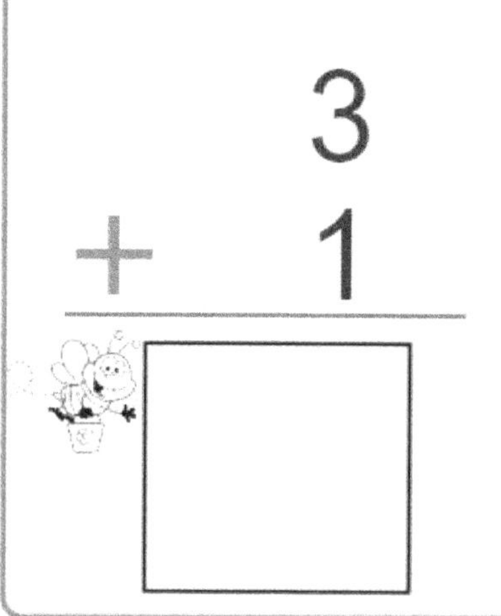
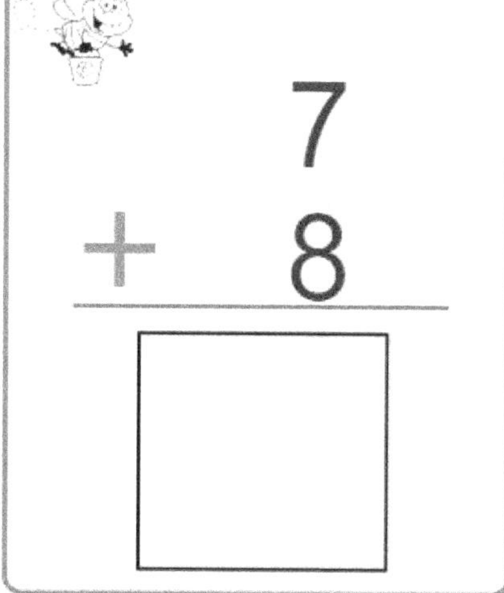

Math Made Easy....

Name : ________________________________

Direction: Count the images. Write the number of images in the boxes above each image and write the total number in the last box.

+

=

+

=

+

=

+

=

Name : _______________________________

Direction: Add the number of images in each box and
write the answer in the last box.

3 + 5

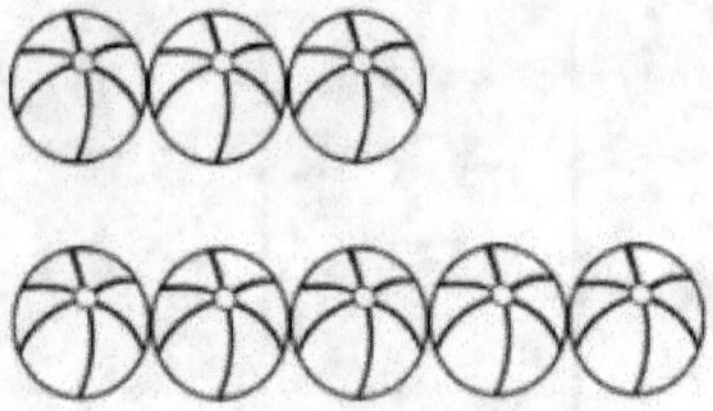

Answer

2 + 5

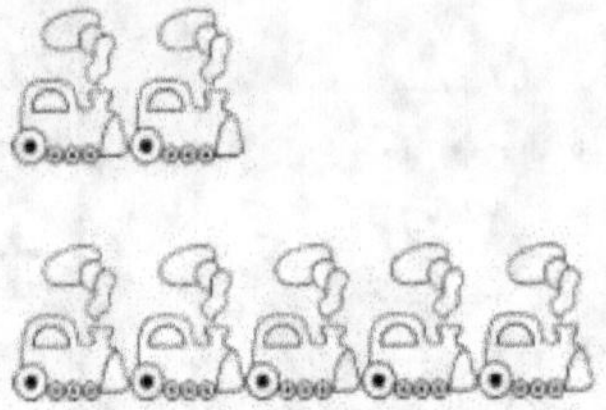

Answer

3 + 1

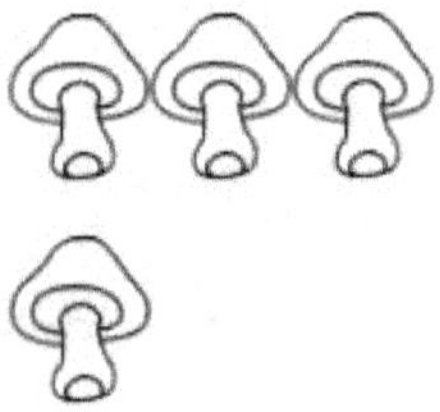

Answer

1 + 2

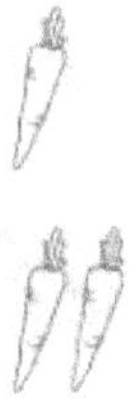

Answer

5 + 4

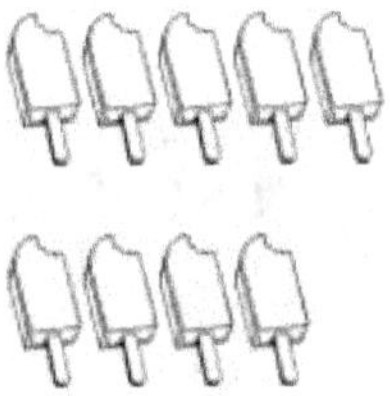

Answer

5 + 2

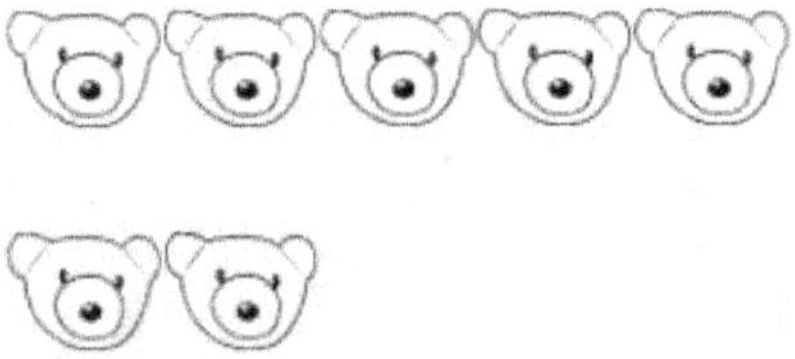

Answer

Name : _______________________________

Direction: Add the images. Draw a line between the total
number of images in each box and the number on the right.

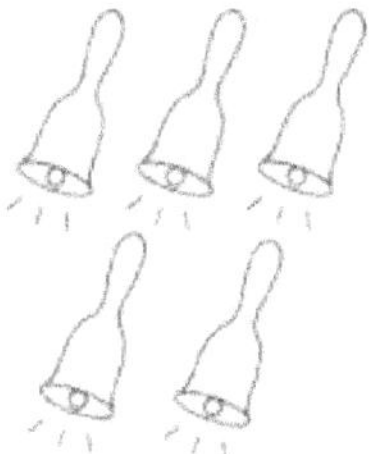

7

5

7

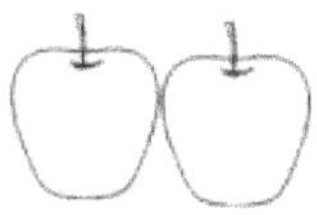

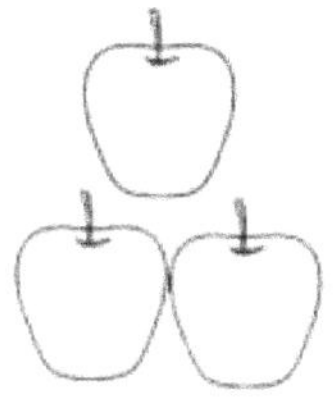

2

Addition Worksheets

	2
+	1

	2
+	8

	10
+	10

	3
+	7

	5
+	4

	8
+	2

	1
+	1

	10
+	7

	10
+	6

Name : _______________________

Direction: Count the images. Write the number of images in the boxes above each image and write the total number in the last box.

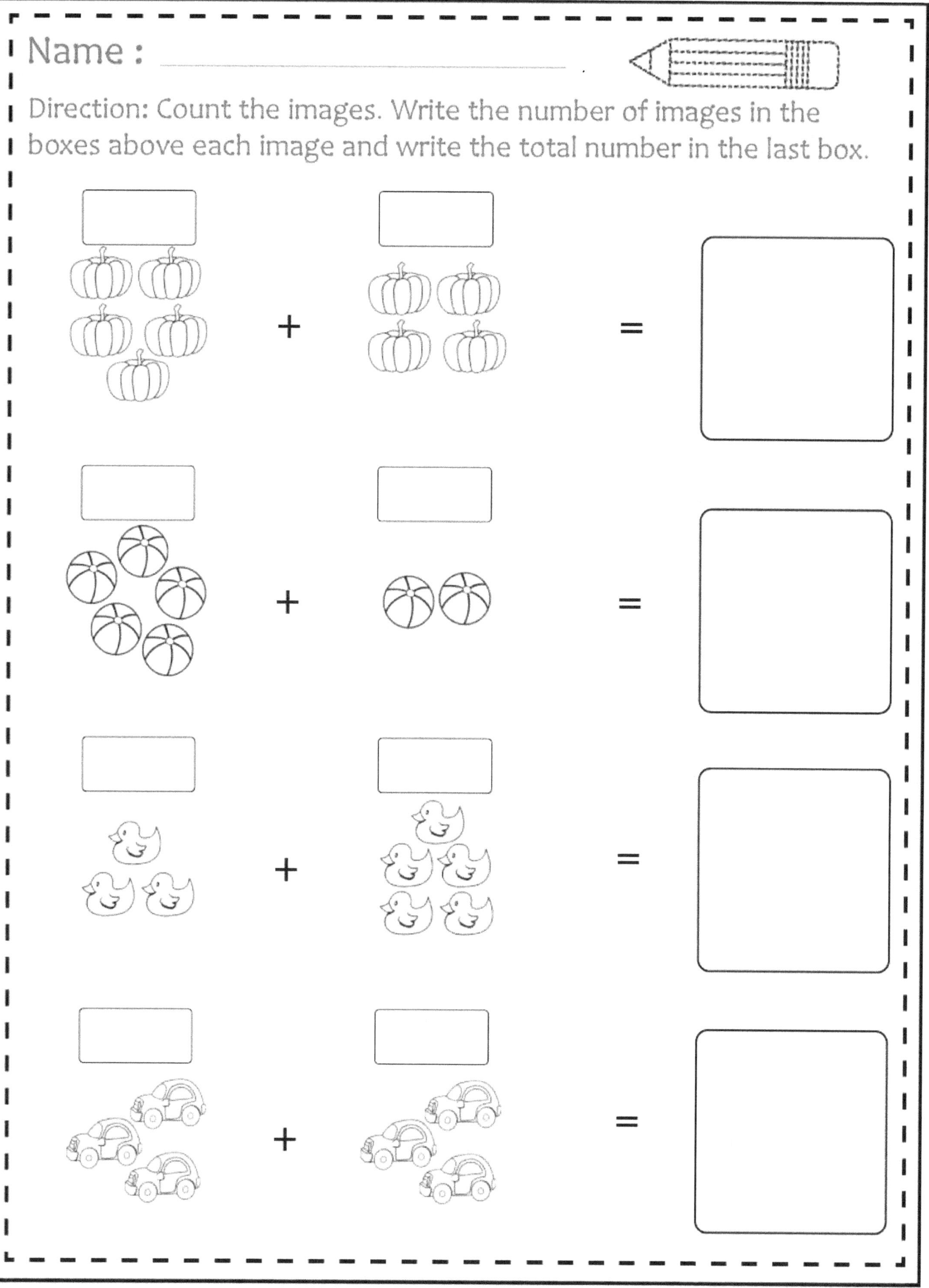

Name : _______________________________

Direction: Add the number of images in each box and
write the answer in the last box.

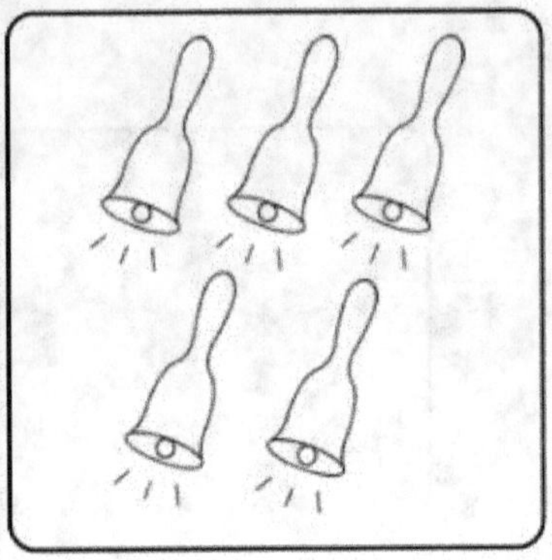

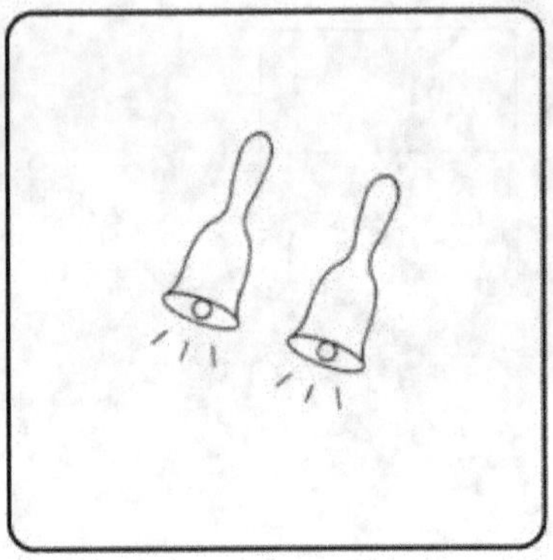

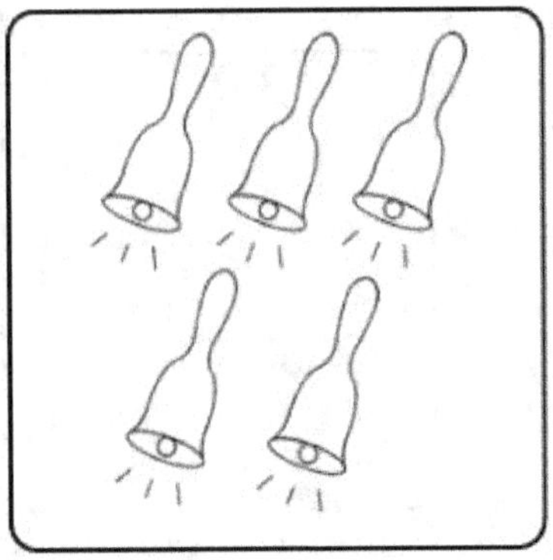

(5 bells)	+	(2 bells)	=	
(1 bear)	+	(4 bears)	=	
(4 gifts)	+	(3 gifts)	=	
(5 bells)	+	(3 bells)	=	

Name : _______________________________

Addition Worksheets

3
+ 5

Answer

4
+ 3

Answer

3
+ 5

Answer

5
+ 2

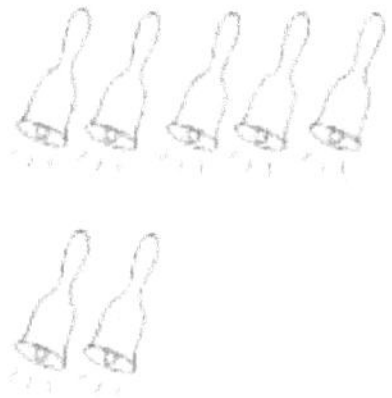

Answer

5
+ 5

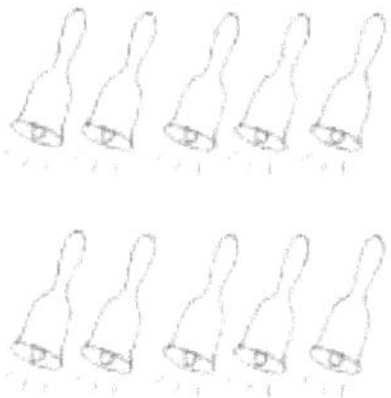

Answer

4
+ 5

Answer

Name : ______________________

Direction: Add the images. Draw a line between the total
number of images in each box and the number on the right.

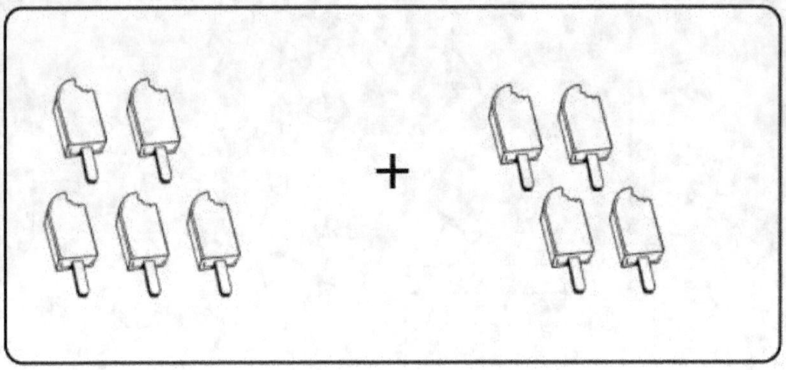

7

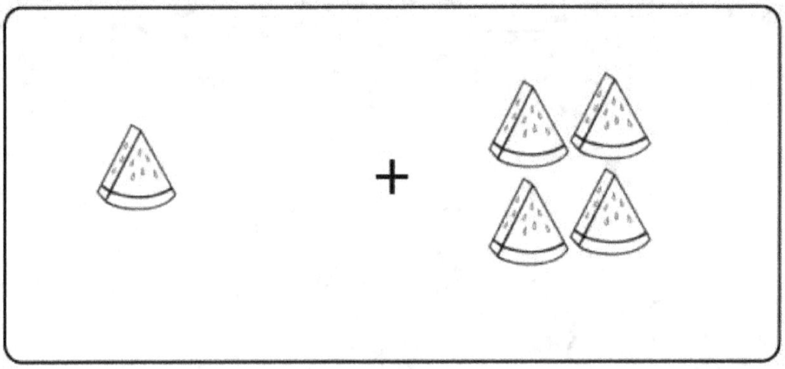

5

9

4

Addition Worksheets 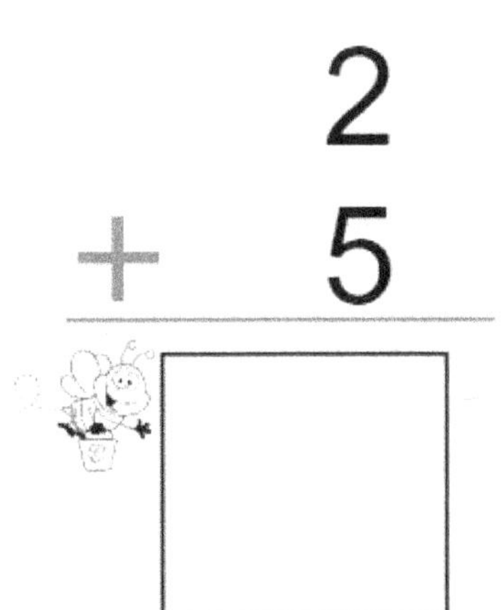

2
+ 5
▢

4
+ 6
▢

10
+ 9
▢

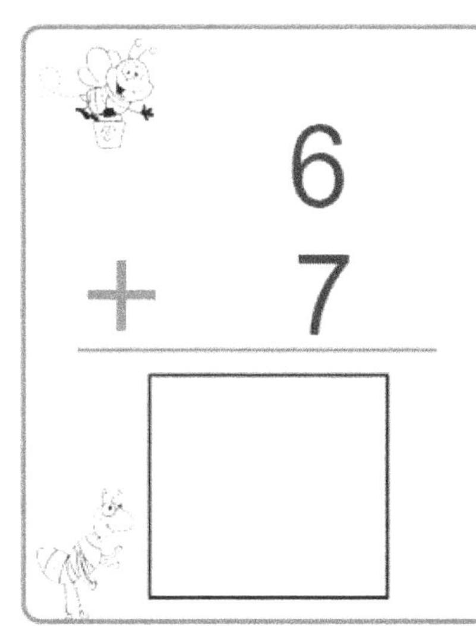
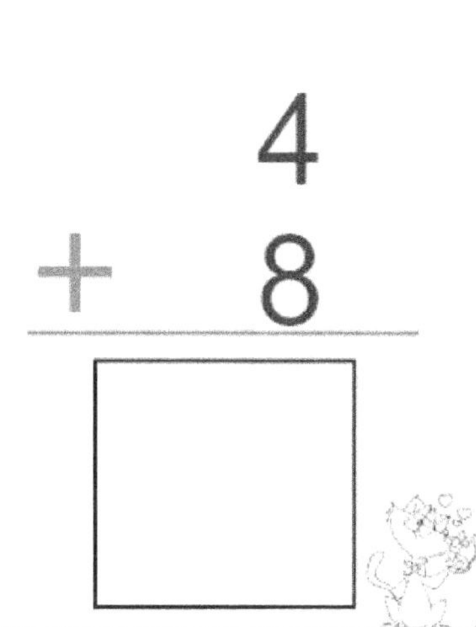

5
+ 1
▢

6
+ 7
▢

4
+ 8
▢

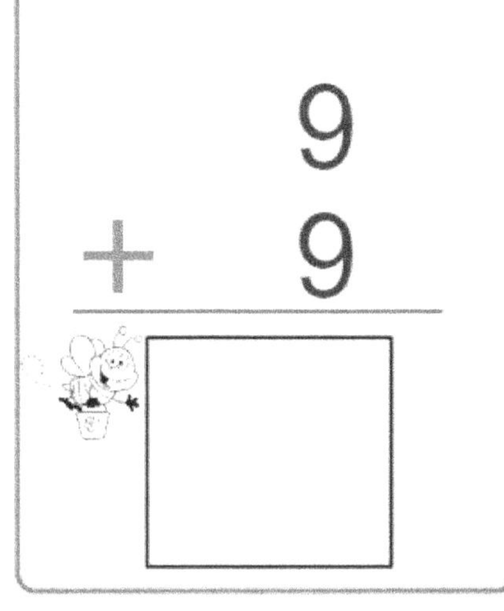
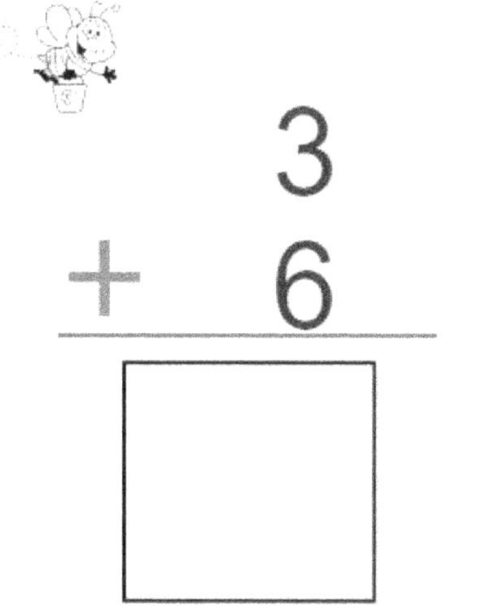

8
+ 1
▢

9
+ 9
▢

3
+ 6
▢

Math Made Easy....

Name : _______________________________

Direction: Count the images. Write the number of images in the
boxes above each image and write the total number in the last box.

Name : ________________________

Direction: Add the number of images in each box and
write the answer in the last box.

Name : ___________________________
Addition Worksheets

3 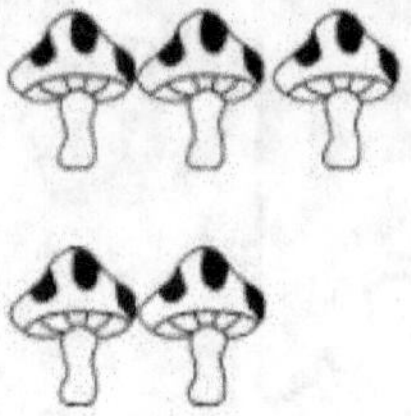
+ 2

Answer

4 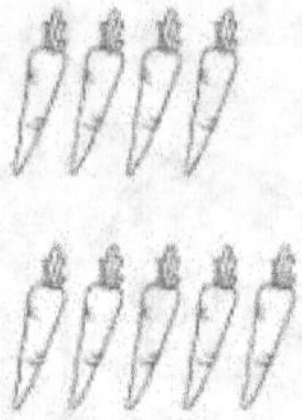
+ 5

Answer

5 
+ 4

Answer

2 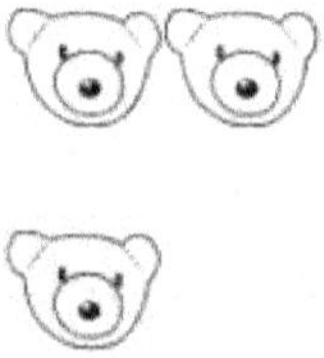
+ 1

Answer

5 
+ 3

Answer

5 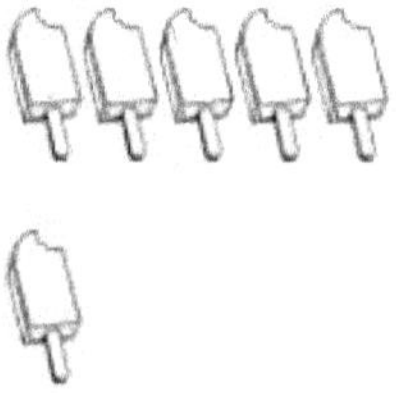
+ 1

Answer

Name : _______________

Direction: Add the images. Draw a line between the total
number of images in each box and the number on the right.

6

4

7

5

Addition Worksheets

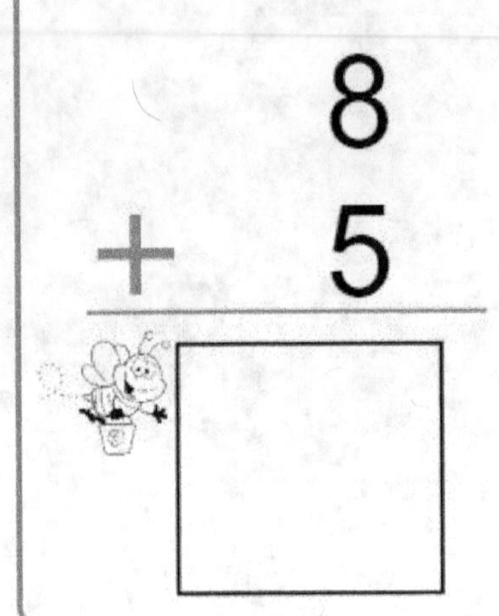

$$8 + 5$$

$$1 + 7$$

$$10 + 6$$

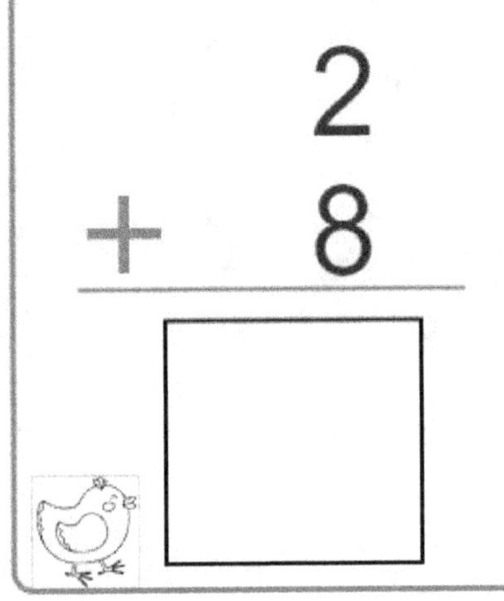

$$2 + 8$$

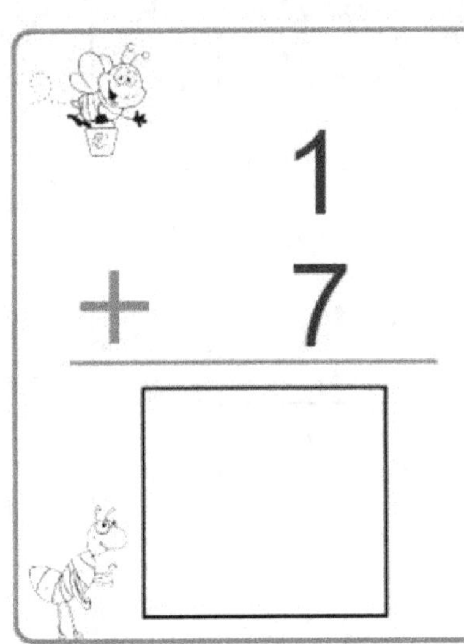

$$1 + 7$$

$$7 + 3$$

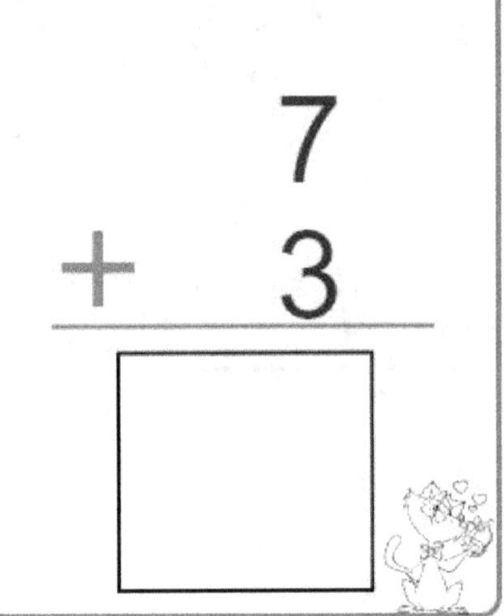

$$5 + 2$$

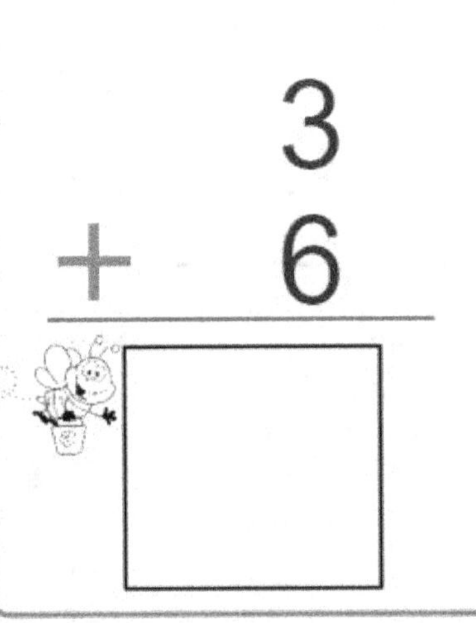

$$3 + 6$$

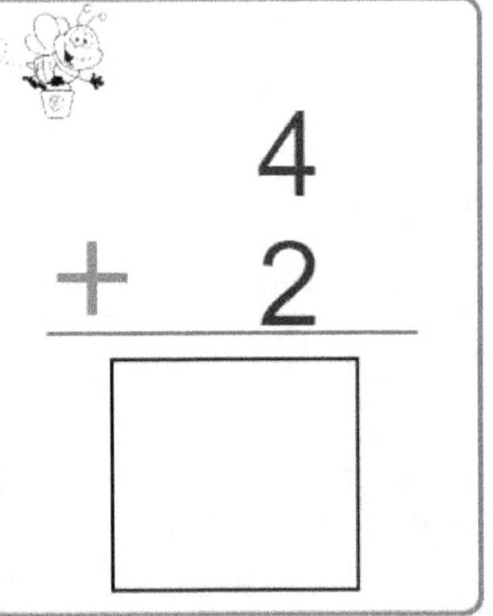

$$4 + 2$$

 Math Made Easy.... 

Name : _______________________

Direction: Count the images. Write the number of images in the boxes above each image and write the total number in the last box.

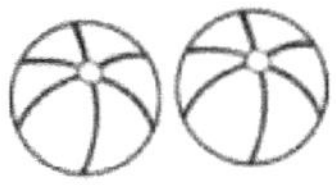

⬜	⬜	
🏀🏀 + 🏀🏀🏀	=	⬜
⬜	⬜	
🐤🐤 + 🐤🐤	=	⬜
⬜	⬜	
🏀 + 🏀🏀🏀🏀	=	⬜
⬜	⬜	
🍉🍉🍉🍉🍉 + 🍉	=	⬜

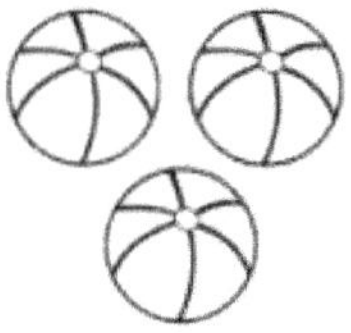

Name : _______________________________

Direction: Add the number of images in each box and
write the answer in the last box.

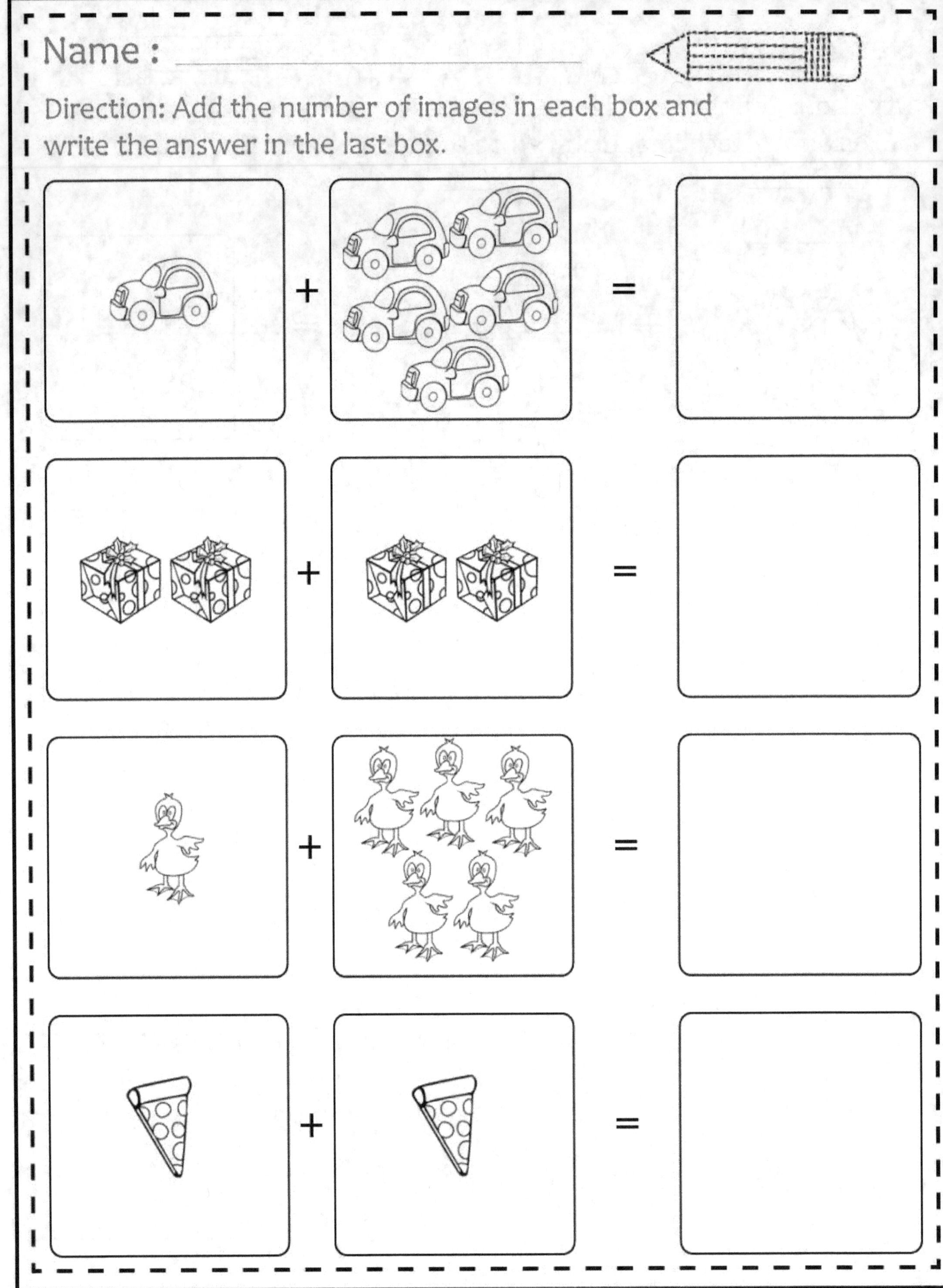

5
+ 5

Answer

5
+ 4

Answer

5
+ 2

Answer

5
+ 1

Answer

3
+ 4

Answer

4
+ 1

Answer

Name : _______________________

Direction: Add the images. Draw a line between the total
number of images in each box and the number on the right.

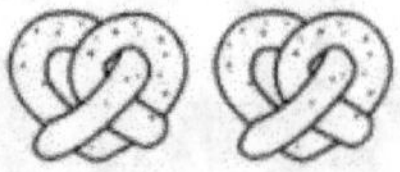

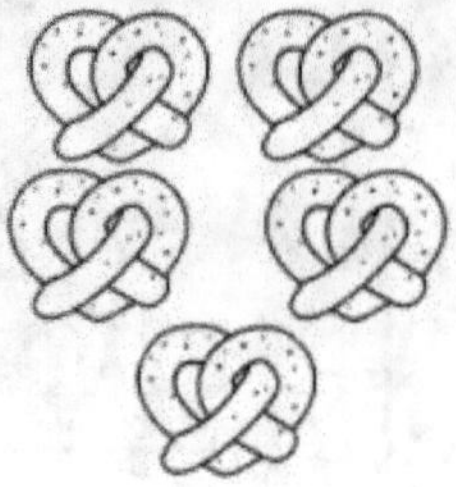

4

3

7

5

Addition Worksheets

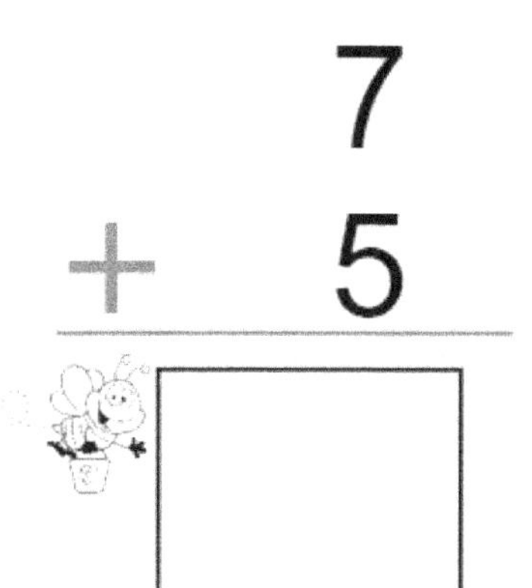

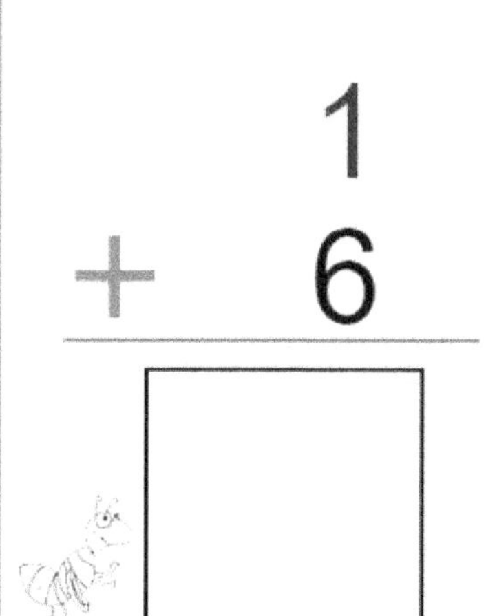

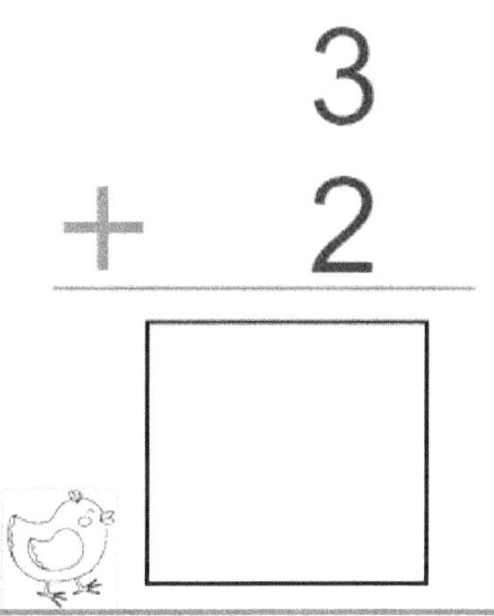

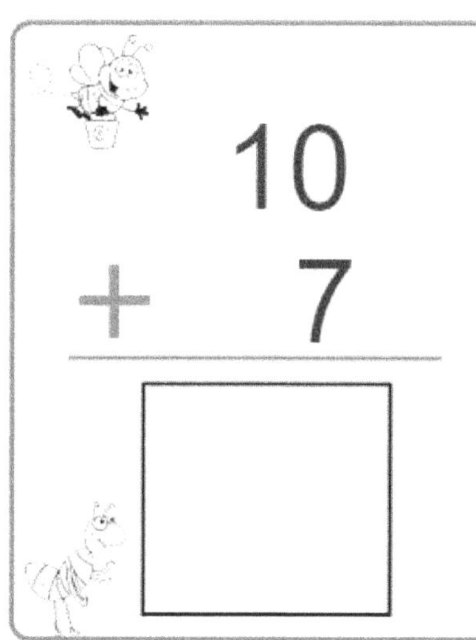

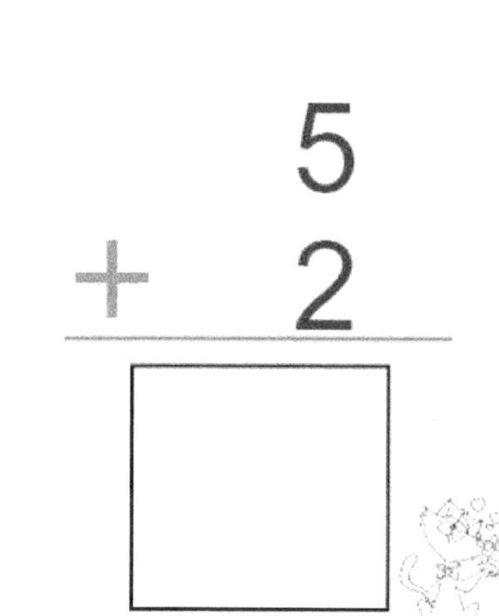

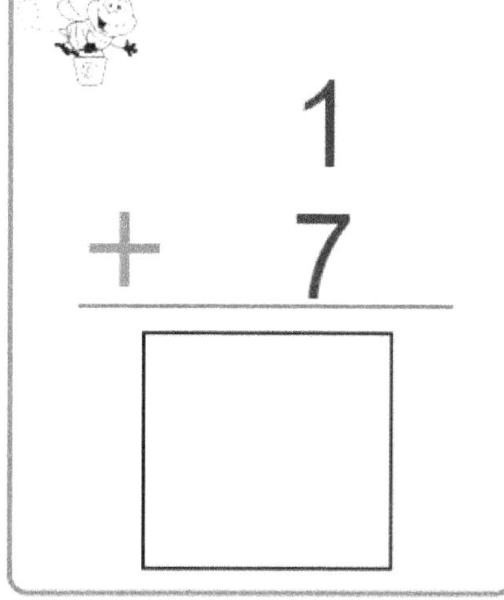

  Math Made Easy....

Name : _______________________________

Direction: Count the images. Write the number of images in the boxes above each image and write the total number in the last box.

Name : _______________

Direction: Add the number of images in each box and
write the answer in the last box.

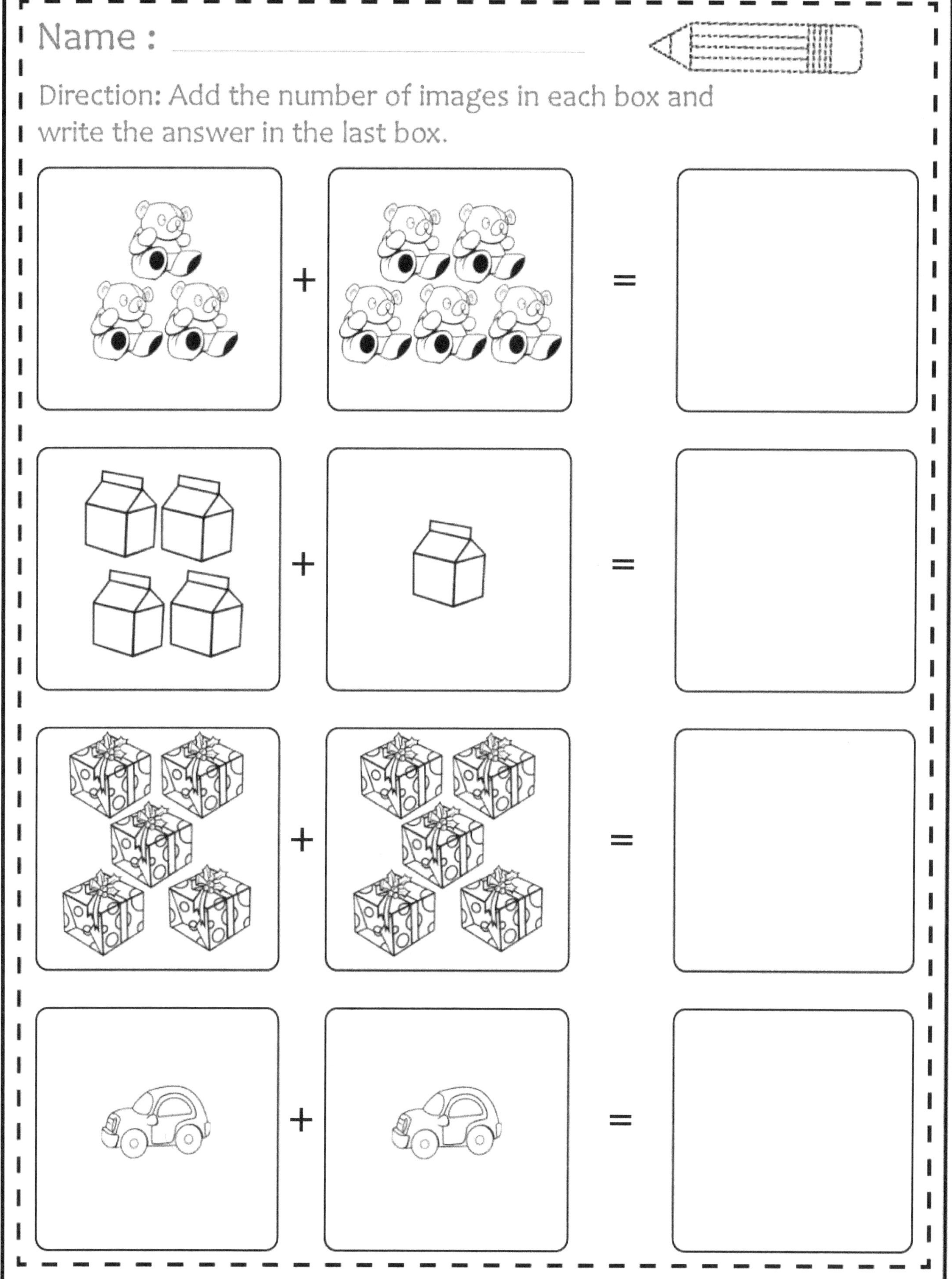

Addition Worksheets

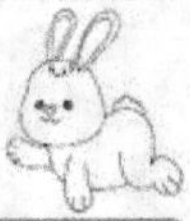

4
+ 4 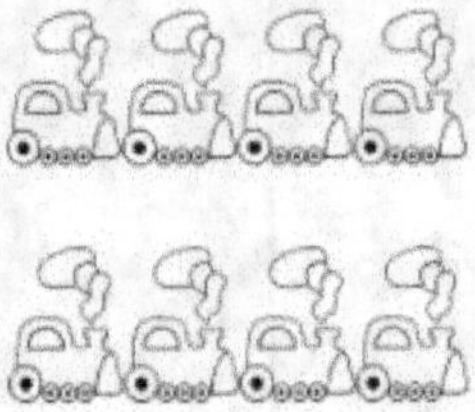
Answer

5
+ 2 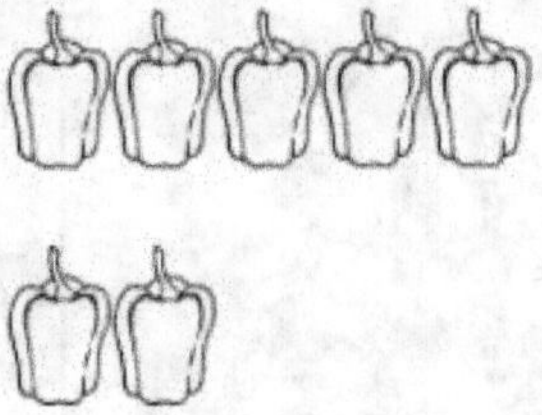
Answer

5
+ 1 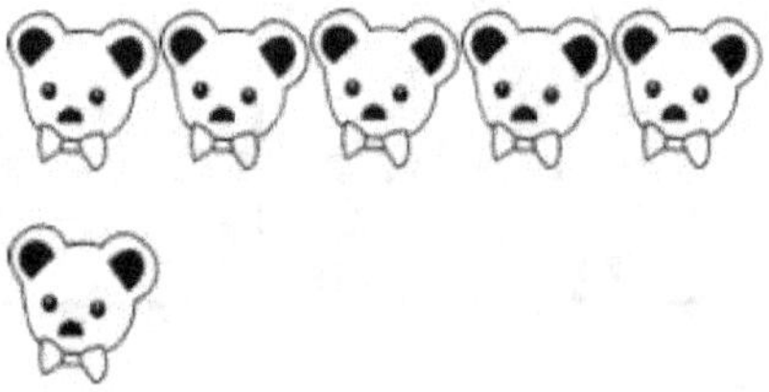
Answer

5
+ 4 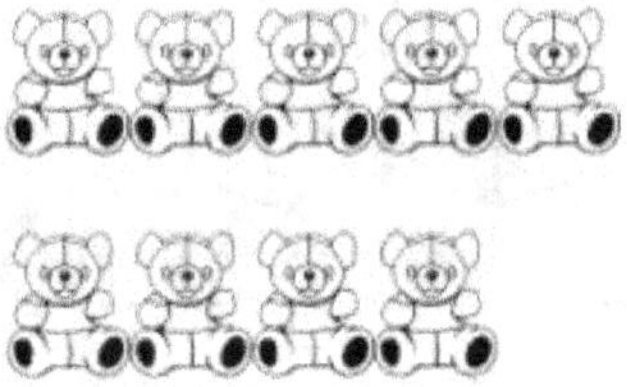
Answer

4
+ 5 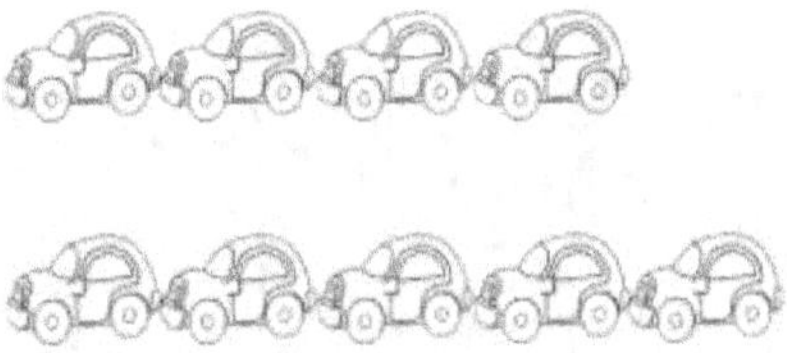
Answer

5
+ 5 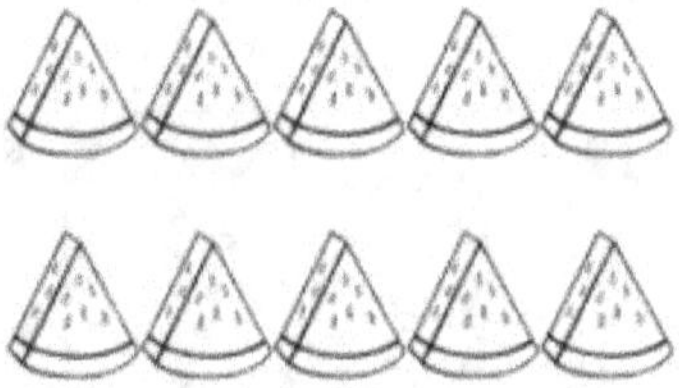
Answer

Name : _______________

Direction: Add the images. Draw a line between the total
number of images in each box and the number on the right.

4

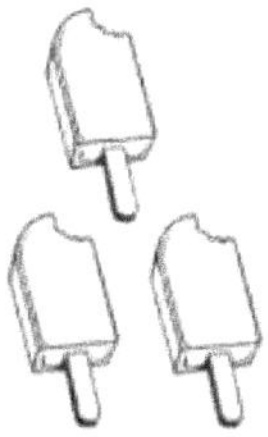

7

4

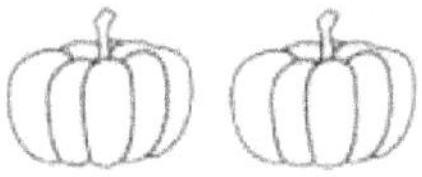

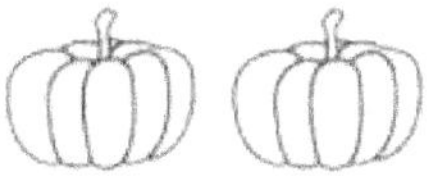

9